Understanding Maps

Charting the Land, Sea, and Sky

Understanding

illustrated by Adolph E. Brotman and with photographs

Maps

Charting the Land, Sea, and Sky

Revised Edition

*by Beulah Tannenbaum
and Myra Stillman*

McGraw-Hill Book Company
New York · Toronto · London · Sydney

Also by Beulah Tannenbaum and Myra Stillman

UNDERSTANDING FOOD
UNDERSTANDING LIGHT
UNDERSTANDING TIME
ISAAC NEWTON

ACKNOWLEDGMENT
The authors wish to thank the men and women of the U. S. Geological Survey and the U. S. Coast and Geodetic Survey for their technical assistance, which was most helpful in the preparation of this book.

The publishers are grateful for permission to reproduce the following: two jacket photographs and pictures on pages 98 and 105, courtesy of ESSA; one jacket photograph and pictures on pages 121 and 122, courtesy of NASA; picture on page 131, courtesy of *The New York Times;* pictures on pages 29, 33, 55, 56, and 58, courtesy of U. S. Geological Survey; pictures on pages 71, 112, 118, 138, 139, 140, and 154, courtesy of Rus Anderson.

Copyright © 1957, 1969 by Beulah Tannenbaum and Myra Stillman. All Rights Reserved. Printed in the United States of America. No part of this publication may be reproduced, stored in a retrieval system, or transmitted, in any form or by any means, electronic, mechanical, photocopying, recording, or otherwise, without the prior written permission of the publisher.

Library of Congress Catalog Card Number: 69-25668

1234567890 VBVB 7543210698

Contents

1. Who Needs Maps? 9
2. Distance, Direction, and Landmarks . . . 15
3. Mapping the Land: Making the Measurements 31
4. Mapping the Land: Drawing the Maps . . 49
5. Mapping the Sea: Finding the Latitude . . 61
6. Mapping the Sea: Finding the Longitude . . 76
7. Mapping the Sea: Nautical Charts . . . 94
8. Mapping the Sky: Man Among the Stars . . 107
9. Mapping the Sky: Weather 123
10. Mapping the Round Earth 133
11. Mapping the Make-believe, the Almost Real, and the Real 146
    Index 157

Understanding Maps
Charting the Land, Sea, and Sky

Who Needs Maps?
1

The Marshall Islands are scattered over more than half of a million square miles of the Pacific Ocean. These low coral islands and atolls form lonely dots of land surrounded by seemingly endless pounding seas. Most of the islands are too far apart to be seen from other islands.

Until recently, the Marshallese people had no written language, no paper or pencils, no compasses, but they had charts, sea maps to guide them from island to island. The people of the islands had so much confidence in these charts that it was not unusual for a fleet of canoes, loaded with men, women, children, and animals, to set out on a voyage to another island hundreds of miles away. Such voyages would take several weeks and sometimes months.

The Marshallese maps, like the one pictured on page 10, are stick charts made of coconut palm or pandanus reeds woven and tied together in carefully planned patterns. Cowrie shells are tied to the reeds. Each shell marks the position of an island or atoll. The lines formed by the reeds show the direction of the waves in the ocean between the islands.

If you have been to an ocean beach, you know that the waves always come in toward the shore. This is true whether the beach faces north, south, east, or west. As a wave approaches an island, it bends or folds around the land mass. Since the opposing wave does the same thing, a wave ridge is built up at the point where the two waves meet. The men who paddled their canoes through the waters around the Marshall Islands had noticed this and learned to judge the position of a distant island by the ridges formed by the waves. They recorded their observations with the reeds in their stick charts, and handed the charts down to their children and grandchildren as a treasured inheritance.

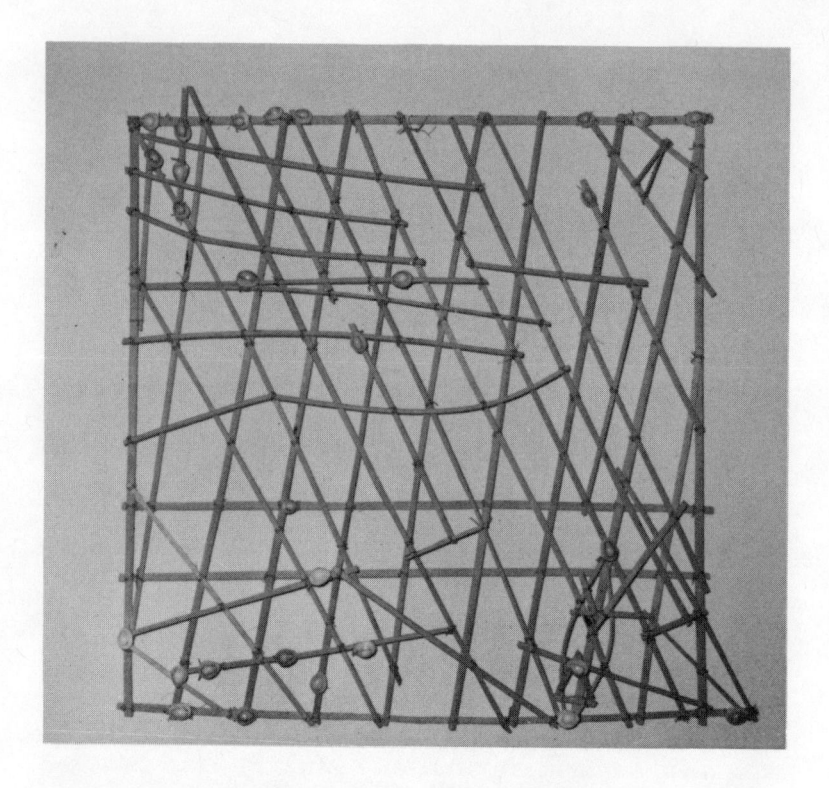

Pacific islanders were not the only people without a written language who used maps. Crude maps drawn in the bare ground often were used by primitive peoples to describe trips or to direct the traveler. Although we may sometimes think that long-distance travel in what is now the United States began with the coming of the Europeans, this is far from true. *Archaeologists* (ar·kee·ah′la·jists), who study ancient cultures, digging in the sites of Indian settlements, have found that even before the time of Columbus, networks of trade routes crisscrossed the continent. Arrows made of flint quarried at Port Ewen, on the bank of the Hudson River, have been found in quantity in the Middle West. Fresh-water pearls from the Mississippi River were widely distributed throughout the continent. And Indians who lived east of the Rocky Mountains knew and traveled the mountain passes that led to the West Coast.

It is easy to understand why travelers need maps. Imagine planning a trip across the United States, or even across your home state, if no maps existed. Or think of a sailor in a modern, fast-moving ship approaching a coast whose shape was unknown and whose reefs or sandbars were uncharted. Even space travelers need maps. Years of work and billions of dollars have been spent mapping the moon in preparation for a manned landing.

However, the use of maps is not limited to travelers. Historians turn to old maps to help them learn about the events of the past. In October, 1957, a handwritten book was discovered in a collection of old books brought from Europe. Included in this book is a map showing part of the mainland of North America. Careful study has shown that the book was written and the map was drawn in the middle of the 1400s, many years before the voyages of Colum-

UNDERSTANDING MAPS

bus. This map, which is called the *Vinland Map,* lends support to those who believe that the first Europeans to land in North America were Norsemen rather than Columbus and his sailors. Even in modern times, the question of who discovered America can stir up heated discussion. Many people were very angry when Yale University released information about the *Vinland Map* just before Columbus Day, 1965.

Of more practical value to modern Americans is the use which engineers make of maps. For example, no major road-building project is planned until accurate *topographical* (tahp·uh·graf′i·cuhl) maps of the area are available. These show the heights and shape of the land; they show hills and valleys. Using such maps, the engineers can study the various routes that the road might take. The maps show which hills could be leveled and gullies filled and where the road would have to curve around obstacles too difficult to remove. The maps show which streams would have to be bridged and which ones would flow through culverts under the roadbed. The maps even show where there are buildings which would have to be moved or torn down to make way for the new road. With the information found on these maps, the road planners can decide on the route, taking into consideration the needs of the travelers for speed and safety, the cost of the construction, and the effect of the new road on the lives of the people who live along the route. In a similar way, maps are used to plan dams, airfields, large industrial sites, and the development of cities.

Geologists (jee·ah′la·jists) use topographical maps in their search for oil and valuable minerals. The heights and shapes of the land are important clues and so are the type and position as well as the folds and tilt of exposed rock. A geologist with a map may not always find the minerals or oil he is looking for, but he is saved the trouble of searching in places where they could never be found.

Students use maps to help them understand the world around them. If there were no maps or globes, how could you "see" the shape of the United States? How could you compare its size and position with those of other countries? Or think of trying to study history without maps. Imagine how many pages would be needed to describe the route of the Lewis and Clark Expedition? And, after you had read all the pages, do you think you would have a clear idea of the extent of the explorers' travels?

Maps also play a big role in helping people enjoy themselves. People who like to watch the night sky often turn to sky charts to help them locate the constellations. Fishermen, hunters, and campers use maps to help them locate desirable sites for their activities. In large cities, people use street maps and bus or subway maps when they want to know how to reach a friend's house or a new shop in an unfamiliar section of the city.

Even people who rarely venture far from home benefit from maps. Many of the things we buy in local stores are carried by large trucks, sometimes all the way across the country. Dispatchers in the company office carefully plan the routes which the trucks will take. They need maps which show not only fast, direct, safe routes but also the heights of the underpasses and how much of a load each bridge can carry. Some of the older underpasses are too low for the larger trucks, and some of the small bridges are not strong enough. When big pieces of machinery or huge rockets for space travel are being moved from where they were manufactured to where they will be used, a truck may be forced to travel far out of its way to avoid a low underpass or a weak bridge. Even though ships and airplanes which carry goods do not face the same problems, the men who operate them also depend on charts to help them reach their destinations safely.

These are only a few of the ways in which maps are used today, and so the answer to the question "Who needs maps?" is "Nearly everyone!" Can you think of anyone who does not benefit from the existence of maps? Perhaps there may be a hermit somewhere who depends entirely on the land around him for everything he needs. But then, even the hermit may have needed a map to find a place to live that was far enough away from all other people.

Distance, Direction, and Landmarks
2

Maps are so much a part of daily life that people seldom stop to think about them. But imagine a world without maps! How would you describe the shape of your home state without drawing a map? Unless you live in Colorado or Wyoming, you might find it difficult. Or think of planning an automobile trip from Atlanta, Georgia, to Seattle, Washington, without a map.

In the early days of automobiles, all you could do was start out in the right general direction and stop at each crossroad to ask your way. Some of the crossroads had markers, many of which were faded by rain and sun and could hardly be read. Sometimes a traveler caught at night on a dark road had to climb a pole and light a match to read a weather-beaten sign. One traveler tells of shinnying up a pole on a wet and stormy night only to find he had left his matches in the car. On the second attempt and three matches later, he finally read the sign. It was an ad for chewing tobacco.

Through the ages, man has found three sets of information which help him reach his destination: *direction, distance,* and *landmarks*. He has learned to use all three in a picture which is called a map.

Today, anyone who wants to plan a trip by car can go to

UNDERSTANDING MAPS

a gas station and get a free road map that shows the three sets of information needed by travelers. Part of the map might look like the one on the next page.

Use this map to plan a trip from New City to Northport. First, you need to know in which direction to travel. Road maps usually place north at the top, but to be sure, look for the *compass rose* on the map; it points to the north. You can see that Northport is northwest of New City.

Next, you need to know how far you will have to travel. You can find the distance by adding the miles that are printed in small unboxed numbers alongside of the road lines. These numbers tell the mileage between such landmarks as cities and crossroads. For example, it is 3 miles from New City to Middle Center and 11 miles from New City to Old Town Corners.

There is no direct road from New City to Northport, but there are two possible routes. If you go in a northerly direction, the first landmark you will reach is Middle Center. Turn left and travel 7 miles past South Lake to Lakeville. Turn left again and continue for 8 miles to Route 3. Turn right and follow Route 3 north 8 miles more to Northport. You could also reach Northport by traveling west from New City for 11 miles on Route 9. Then, turn right at Old Town Corners and Follow Route 3 north for 17 miles to Northport.

The map can help you decide which road to take. If you start out in a northerly direction, the distance is 26 miles. The scenery probably will be interesting because you will drive along the shore of South Lake, but you will have to go through two villages. The road from Lakeville to Route

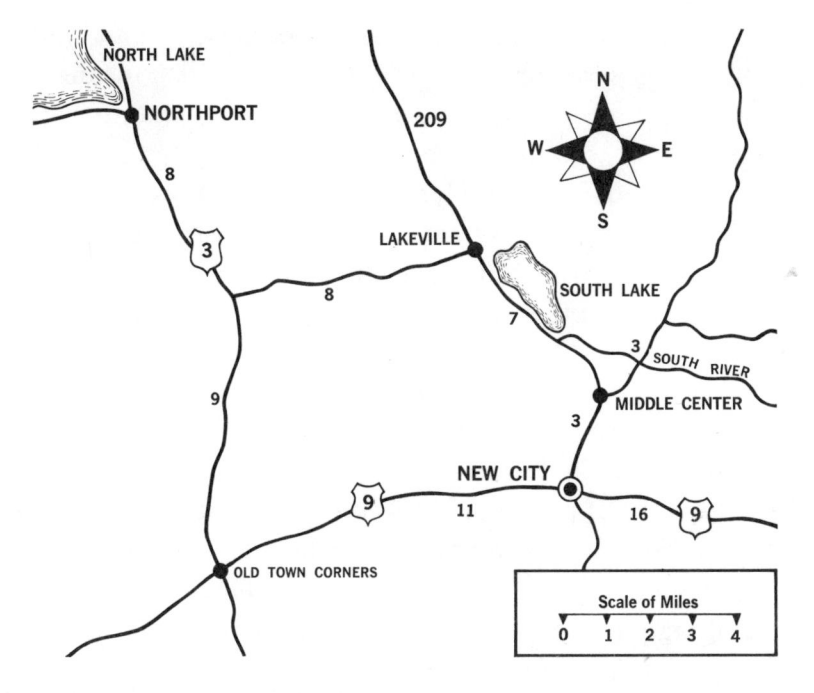

3 is not a main road, for it has no route number. It probably will not be as good a road as those which are numbered. If you choose the other route, the distance is 28 miles, and you will go through only one village. There are no landmarks like South Lake to give you a clue to the scenery, but you know that you will travel on good roads, for both of them are numbered highways.

It is simple enough with this road map to plan your trip from New City to Northport. With the information found on road maps, it would not be too difficult for you to plan a trip across the United States.

But how does the map-maker collect this information? How does he measure the distance and determine direction?

You can read distance from a map because you and the map-maker agree on a unit of measure. An *inch* means a definite distance to both of you. Because you use a ruler so often, you know there are 12 inches in a *foot* without having to figure it out each time. You also know that there are 3 feet in a *yard*. And even though you probably have never measured a mile yourself, you have learned that there are 5,280 feet in a *mile*. Today the unit of measure which we call a "foot" covers the same distance whether it is used in England, Florida, or Alaska. This was not always true. One hundred fifty years ago in Europe, the foot measure came in 280 different sizes.

The story of how man worked out fixed measurements covers thousands of years. The earliest units were based on parts of the human body. The length of the forearm, the marching pace of a soldier, and even the size of a king's foot were used as units of measurement. You can imagine the confusion that occurred when the old king died and the new king's foot became the standard.

It was not until 1824, when the British Parliament set up a new standard, that the present length of the yard was finally accepted. The modern standard ruler is kept in London. It is made of brass and has a gold button at each end. The distance between the buttons is exactly one yard.

While the English were deciding the exact length of the yard, the people of France were working out another way of measuring distance. The French system is based on the *meter*. A meter is a little longer than a yard. It is 39.37 inches.

The French wanted their measures to depend on something which would never change. Instead of using a part

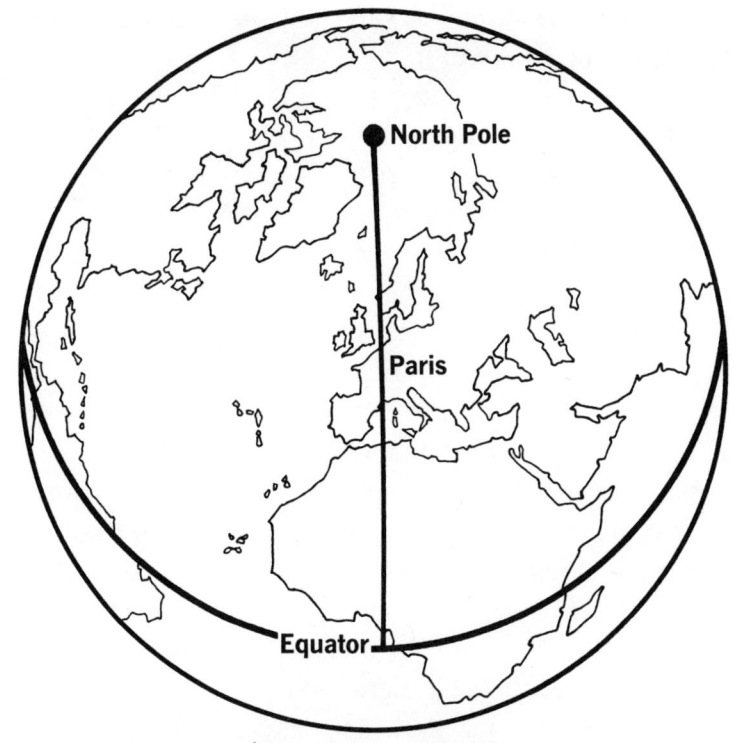

THE PARIS MERIDIAN

of the human body, they decided to use the earth itself. In 1791, the Paris Academy of Sciences defined the meter as one ten-millionth of the length of a line starting at the North Pole, running through Dunkirk, Paris, and Barcelona, and ending at the equator.

In 1960, the 11th General Conference on Weights and Measures, which met in Paris, set up a more accurate standard for the meter. They chose the wave-length of the gas Krypton 86 under special conditions.

It took thousands of years of experimenting for man to arrive at fixed standards of measurement. But with this

UNDERSTANDING MAPS

knowledge the map-maker can show distance on the map so that anyone can read it.

Distance on a road map can be shown in two ways. You may add the small unboxed numbers as you did in finding the distance from New City to Northport, or you may figure the distance by using the *scale,* which is part of all maps. The scale tells you how many miles there are in each inch on the map.

Determining the scale is one of the first things a map-maker does. He must decide the size of the finished map and he must know the size of the area he is mapping. To find the scale, he divides the size of the area by the size of the map. The answer is the number of miles of land which will be represented by each inch on the map. For example, if you are making a map 100 inches wide of an area which is 1,200 miles wide, the scale would be 1,200 miles divided by 100 inches, or 1 inch represents 12 miles.

If you look at the next picture, you will see that, while the maps are the same size, the scales are very different. Because the United States is so much larger, the scale of its map is 1 inch represents 480 miles, while that of Connecticut is only 1 inch represents 28 miles.

Knowing distance is not enough; the traveler also has to know direction. Many birds have an amazing sense of direction, and, without maps or instruments, can find their way easily over vast sea and land distances. For example, some North American swallows fly all the way to Argentina each winter, and the next spring are often found again in their old nesting places.

Man has no such ability, but direction is as important to a man who wishes to travel as it is to birds. Earliest man

MAP OF THE UNITED STATES Scale 1 inch = 480 miles

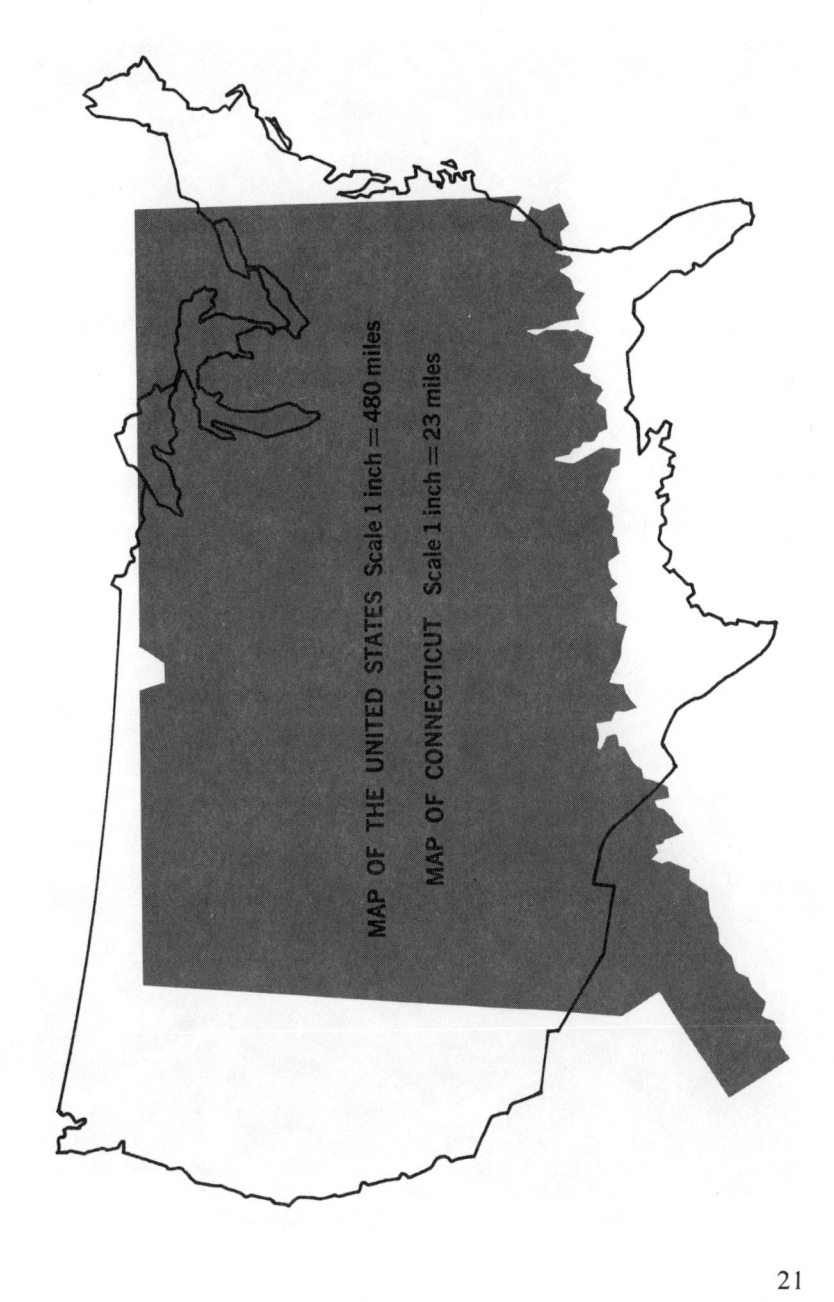

MAP OF CONNECTICUT Scale 1 inch = 23 miles

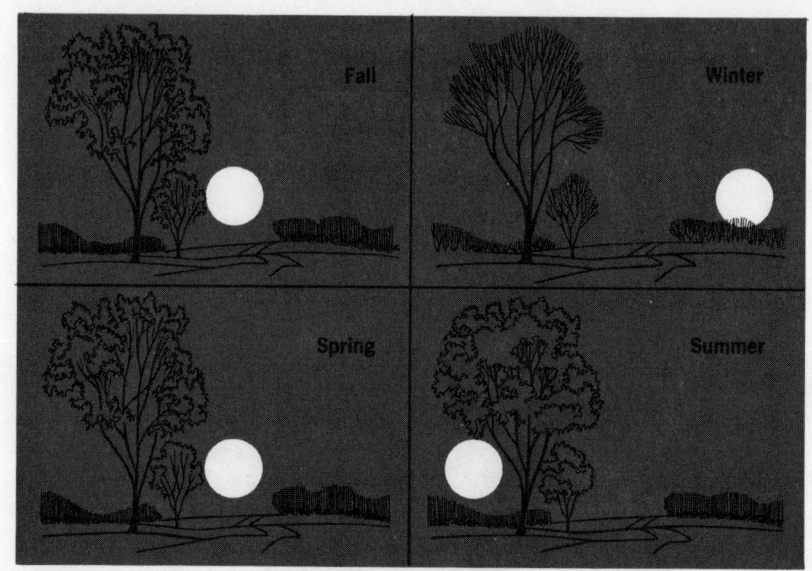

POSITION OF SUNRISE

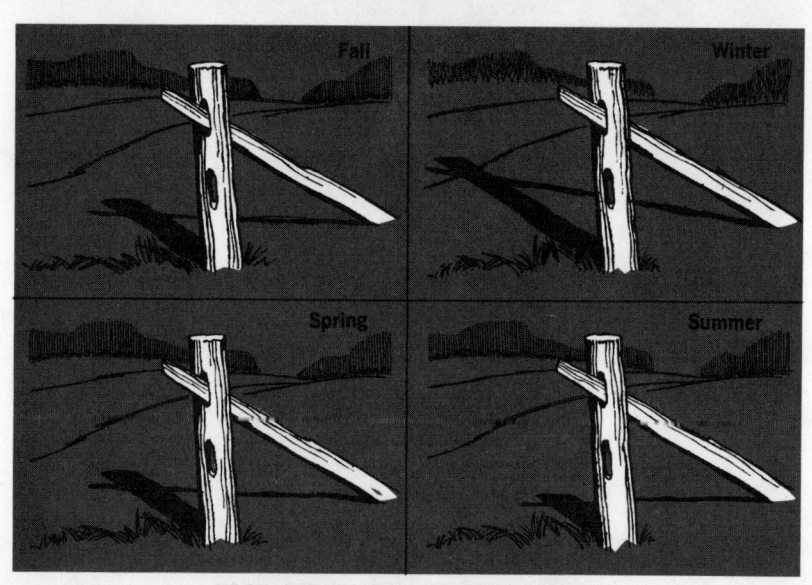

NOON DAY SHADOW POINTING NORTH

used the rising sun to tell direction. We say that the sun rises in the east, but this is not exactly true. As the earth moves around the sun, the position in which the sun rises changes slightly each day. In the Northern Hemisphere, summer sunrises occur farther toward the north than winter sunrises. However, no matter where the sun appears to rise, one thing is always true: A stick put into the ground so that it stands straight up will cast its shortest shadow at noon. In all parts of continental United States, this noonday shadow will point due north every day in the year. In Hawaii, Puerto Rico, the Virgin Islands, and some of the Pacific Islands such as Guam, the noon shadow is due south for a few days each summer because, on those days, the noonday sun is north of the islands.

This way of finding direction is helpful on land, but imagine a sailor trying to find north in this way while his ship tosses about on a rough sea. When man decided to travel across great stretches of open water, he needed another guide.

While watching the heavens, the men of ancient times who lived in the Northern Hemisphere discovered a strange fact. As the stars wheel around in the night sky, one star always seems to be motionless. No matter where you stand, whether it is outside your house, or at the other end of town, or in Paris, or Vladivostok, or Alaska, this star is always north of you. It is the *North* or *Pole Star*.

NOW TRY THIS

You can find the Pole Star very easily. On a clear night, look toward the north for the Big Dipper. It is made of

THE BIG DIPPER

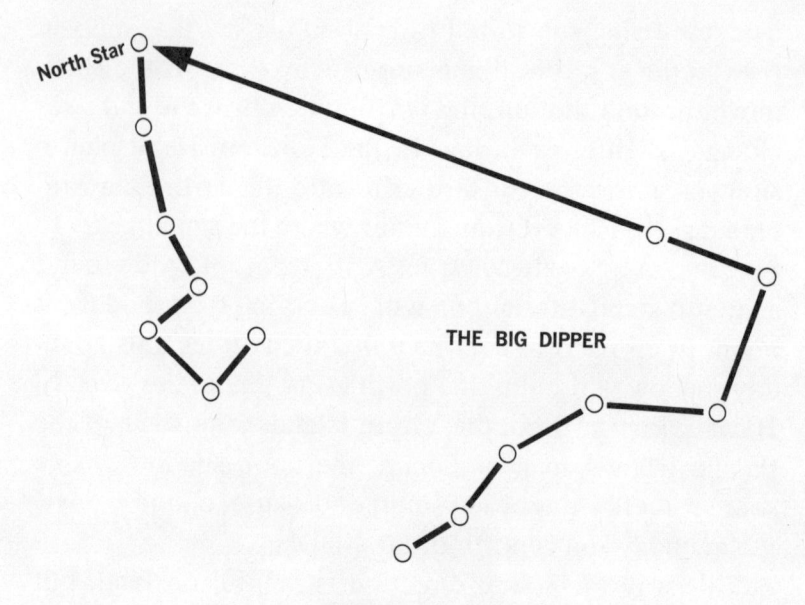

FINDING THE POLE STAR

seven stars. The two stars in the cup which are farthest from the handle are called the pointers. If you could draw a line through these stars, it would lead to the last star in the handle of the Little Dipper. This star, which is not as bright as the pointers, is the North Star.

When the sky was cloudy, neither the sun nor the stars could help the sailor. What man needed was something he could use night or day, rain or shine, on land or on sea.

The answer to this problem was the *compass*. No one knows who invented the compass. The honor has been claimed by many nations, and it is probable that there were many different inventors. We do know that in the Ural Mountains of eastern Europe, a very strange kind of stone is found. Many people thought it was a magic stone for two reasons. First, it could pull iron toward itself. Second, if a thin piece were mounted so that it could move freely, it would point toward the north. It was given the name *lodestone* (load'stone), or lead stone, because it could lead travelers by pointing direction. Today it is called *magnetite* (mag'nuh·tight). A little magnetite is found with other iron ores in many places on earth. There is a large deposit of magnetite in Magnetic Cove, Arkansas.

No one knows exactly what magnetism is, but for more than 300 years scientists have been studying magnets and have discovered many facts about them. They have found that the magnetism is strongest at the ends, or *poles,* of a magnet, and that if you break a magnet in half, each part will be a magnet with two poles. If you break the two new magnets in half again, you will have four magnets, and so on down to the smallest piece you can get. A magnet has two unlike ends: a "north" end and a "south" end. If you place two pieces of magnetite close to each other with opposite poles facing, they will attract each other. If like ends are placed close together, they will repel each other.

The earth also acts like a huge magnet, with one magnetic pole presently near Prince of Wales Island in northern Canada and the other in Wilkes Land in Antarctica. And so a compass needle will point north because its "north-seeking" end is attracted to the earth's north magnetic pole.

The earliest compasses were made by fastening a sliver of lodestone shaped like a needle to a reed. This was then floated in a bowl of water. During the Middle Ages, trade between Europe and the Orient became important. The

UNDERSTANDING MAPS

kings and knights wanted rich silks to wear and spices for their foods. The trade routes across the great deserts of Asia were traveled by camel caravans. Leading most of these caravans were the owners of such compasses. They kept their lodestones a carefully hidden secret, so that they were looked upon as great magicians. It was so difficult to cross the great deserts without this "magic" that the guides were able to charge very high prices for their services.

NOW TRY THIS

You can make a compass similar to the ancient compasses by stroking a darning needle with a magnet. You must be sure to stroke the needle in only one direction for about three minutes. Always use the same end of the magnet for each stroke. When the needle attracts a paper clip, it is ready for use. Run the needle through a small piece of cork and then float it in a bowl of water. Be sure that there are no iron or steel objects nearby so that one end of the needle will point to the magnetic north.

For less than one dollar, you can buy a good compass. It will have a magnetized needle with the north-seeking end probably colored blue. The needle will be mounted on a pivot so that it can swing freely. The compass case will have a glass top, and at the bottom, there will be a compass card with the directions printed on it.

Today, most maps have a symbol, the compass rose, to help you find direction. Suppose you are stranded in the middle of a strange crossroad and you do not know which of the four roads will take you home. If you have a map and a compass, it will be easy to find your way. First find north with your compass. Then hold the map so that the arrow on the compass rose points north. Now you can see which of the four roads will lead you home. When you use a map and compass this way, you are *orienting* the map.

If you do not have a compass, you may use landmarks to help you find direction. A *landmark* can be anything that is used as a guide. For example, if you came out of the woods on the road between Middle Center and Lakeville and did not know which way to go, you could use the map on page 17 to help you. If South Lake is on your right as you follow the road, you know that you must be heading northwest. If it is on your left, you are traveling southeast.

Since primitive times, landmarks have been used in map-making. If man could travel like a bird in a straight line, he would need to know only direction and distance. But imagine what might happen if you wanted to tell a friend how to find his way to your home. If you told him to go straight east for one-half mile, he might find himself jumping over fences, climbing roof tops, and even swimming a pond. On land, direction and distance are not always enough; landmarks are helpful.

Landmarks were important also in the history of sea travel. Before the invention of the compass, sailors tried to stay within sight of land. They learned to know each high cliff and church steeple. This kind of sailing was called church steeple navigation and the guides were called landmarks.

Many old maps have pictures of landmarks on them such as the ones in the first illustration on page 28. Usually, road maps do not have these pictures, but they do

Airports

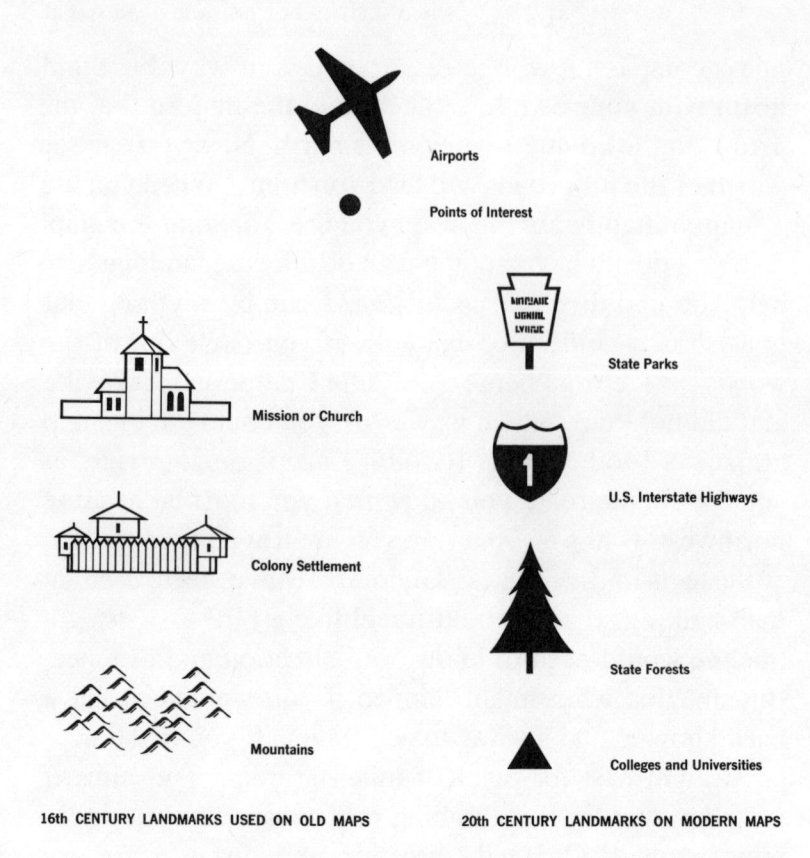

Points of Interest

State Parks

Mission or Church

U.S. Interstate Highways

Colony Settlement

State Forests

Mountains

Colleges and Universities

16th CENTURY LANDMARKS USED ON OLD MAPS   **20th CENTURY LANDMARKS ON MODERN MAPS**

show landmarks. Each city marked on the map, each lake, river, or railroad crossing is a landmark for today's traveler. Some modern maps use certain symbols to show landmarks, such as those in the second illustration above.

You probably have many of your own special landmarks that help you find places. Perhaps you always turn left at "that yellow gas station" when you visit your grandparents. Or, you may always turn right at the drug store on your way to the library. In this case, the "yellow gas station" and the drug store are your landmarks.

DISTANCE, DIRECTION, AND LANDMARKS

The trouble with landmarks is that they may disappear. A house or a store may burn down; small railroads sometimes stop running and their tracks are ripped up for scrap metal; trees die and rot away. Even a stream may dry up or change its course. While landmarks are an important part of map-making, they are less dependable than direction and distance.

Disappearing landmarks often cause trouble to land owners and the *surveyors* who measure the land. Most old deeds use landmarks to describe property. Such a deed may read ". . . bounded northerly by a line beginning at the mouth of Lewis Creek or Kill and running thence south 85 degrees east to a large oak tree." After 150 years, the oak tree and its stump have disappeared and the mouth of the creek may have moved many yards.

Because many natural landmarks change easily, the surveyors who make official maps of the United States have their own landmarks. These are called *bench marks*. A bench mark is usually a bronze plaque set in rock or concrete. It is shown on a map by an X followed by the initials B.M.

UNDERSTANDING MAPS

Fixed standards of measurement, methods of determining direction, and an understanding of the use of landmarks are the tools which the modern map-maker has inherited from the past. With this information, you could map the land on which your house is built. By measuring the distance between landmarks with a tape measure and finding the direction with a compass, you could make an accurate map. The map-maker, however, cannot walk from landmark to landmark with a tape measuring continents and oceans, but he has found ways to map them.

Mapping the Land: Making the Measurements
3

No one can predict where exciting discoveries will be made. A clerk working over a pile of figures in the office of the Indian Trigonometric Survey discovered the highest mountain in the world. On the chart he was using, the mountain was called simply Peak XV. The people who lived within sight of the mountain called it Chomolungma (cho'mo·lung'mah), and you call it Mt. Everest. How could a discovery of such importance be made by a clerk who may never even have seen the mountain?

This can happen because map-makers are divided into two separate groups. One group works in the area that is being mapped. These are the field workers who collect the necessary information. The other group usually works far from the area that is being mapped. These people use the information collected by the field workers to plot the actual maps.

The men in the field who supplied the information that led to the discovery of the earth's highest mountain worked under most difficult conditions. Their instruments were not as good as those which are used today. They had no airplanes to help them, and they could not

UNDERSTANDING MAPS

enter the two countries at the base of Mt. Everest. Tibet, on the north, was closed to travelers until 1920, and Nepal, on the south, would not permit anyone to enter until after World War II. In spite of these problems, the results of their calculations were very close to the actual height of the mountain.

The men who survey the land for map-making are called *geodetic* (gee′oh·det′ik) surveyors. Their work is most exciting and adventurous. Maybe we do not read about them in the newspapers as often as we read about famous mountain-climbing expeditions and jungle safaris, but often the surveyor has explored the area before the "adventurer" arrives. Once, in a Central American country, a party of climbers announced with great fanfare that they were planning to be the first to climb a high mountain in one of the provinces. The climb was successful, but it was never reported in the newspapers, for, when they reached the peak, they found the bronze bench mark of the Inter-American Geodetic Survey set in the bedrock. The surveyors had been there before them!

The United States Geological Survey is the government agency which prepares the basic land maps of the United States. The Geological Survey has divided the country into 6,400 sections called quadrangles. Quadrangle maps like the one on page 33 show distances, directions, and the heights of the land. Contour lines are used on these maps to show the shapes of the hills and valleys. Quadrangle maps include such natural landmarks as rivers, ponds, and swamps, and also man-made landmarks like roads, houses, public buildings, parks, and mines, even old ones which have been abandoned. Quadrangle maps of the

Idaho Springs quadrangle

area where you live can usually be purchased at large book or stationery stores, or you can order one from the U.S. Geological Survey, Washington, D.C. 20242.

Geological Survey maps are public property, and so can be copied by anyone. Since it would not be practical for each private map-maker to send a crew of field men out to survey an area every time a new map is made, the map publishers usually base their maps on the government quadrangle maps. The private map-maker combines the quadrangles and adds or removes details, depending on the way the map is to be used.

Perhaps you think that all the world has been mapped and that there is not much left for the field men to do, but this is not the case. Even the United States is not yet completely mapped. Although there is some kind of map of all its parts, you may live in a section of the United States for which there is not yet a U.S. Geological Survey quadrangle map. The plan is that by 1981 a modern map will be available for each of the quadrangles. Then it will be necessary to start over again, to keep the maps up-to-date. New roads and buildings will have to be added, and contour lines will have to be changed where bulldozers have been at work.

Map-making is an ever-continuing process. For example, in 1867, when the United States was purchasing Alaska from Russia, a map-making survey was begun. One hundred years later, the task of mapping the forty-ninth state with its 586,000 square miles of land still goes on. Throughout the years, fleets of survey ships on six-month expeditions have cruised along the 34,000 miles of coastline gathering information. Some of the areas that once were considered mapped had to be remapped when better instruments were made. When ways to penetrate fog became available after World War II, it was found that some islands were as much as six miles from their charted positions. Even a fraction of a mile is no small error to the pilot of a ship sailing along that rocky, fog-bound coast.

If there is much work ahead in the United States, there is even more in the rest of the world. If you want to help with this work when you are older, there will be many unsurveyed lands waiting for you. Some areas of Asia, Africa, and South America have never been mapped. In 1891, a plan for a universal map was first discussed at a meeting in Berne, Switzerland. By 1913, it was agreed that the nations of the world would work together on such a map. In 1962, another conference was called. Although the Communist countries did not attend, other nations promised to push the project as rapidly as possible. If the map is ever finished, it will be made up of 2,000 sheets or separate maps. All will have a scale of 1,000,000 : 1 which means that 1 kilometer on the earth's surface will be shown as 1 millimeter ($1/_{1000}$ meter) on the map.

It is not only lack of cooperation which holds up the universal map. There are so many problems in measuring the land that man has had to invent many ways of doing it. One method is actually to measure the ground with some kind of ruler. For example, many centuries ago the Egyptians had an unusual problem. Each year the Nile River flooded its banks and washed away the landmarks which divided the fields. After the flood had passed, it was necessary to remeasure all this rich valley land. In order to make the job easier, and be sure everyone had his fair share and paid the right taxes, the Egyptians used ropes of exactly the same length to measure the fields each year. Today, in the pyramids, which are the tombs of the ancient Egyptian kings, you can see pictures of the men at work measuring the land.

When George Washington was employed to survey the American wilderness, a standard surveyor's chain 66 feet long was used. A steel tape 100 or 200 feet in length replaced the chain. Finally, a 160-foot *invar* tape made of nickel and steel was used. Neither stretching nor temperature changes have much effect on such a tape. This tape method is accurate for measuring short, level distances, but most of the distances which the surveyors must measure are neither short nor level.

NOW TRY THIS

To see for yourself how much difference levelness makes, try an "ant's-eye view" of a landscape. Put two sticks into the ground and make "ant" mountains by placing stones or boxes between them. Measure the distance between the sticks with a tape measure. (If you do not have a tape, a piece of string will do.) Be sure that the tape goes up and down over the tops and sides of the boxes or stones. Now remove the "mountains" and measure the distance between the sticks again. You may be surprised to discover how great the difference is.

It was easy enough for you to move the stones or boxes, but the geodetic surveyors cannot move mountains and valleys. They need more than a tape measure to find the exact distance between two places on the earth's surface.

The shape of the earth also causes trouble for the surveyors. Because it is round, a traveler who wishes to go in a straight line from New York to Chicago would have to tunnel through the earth. Since he cannot do this easily, he has to travel on the surface and he needs to know the distance along a line which follows the earth's curve.

Since the earth is so large, the amount of curve at any one place is slight, and in measuring very short distances, surveyors can act as if the earth were flat. In measuring greater distances, the curve of the earth becomes very important. In the diagram you can see what happens when there is a big curve between two points. The distance between *A* and *B* on the straight line is 3 inches. On the curved line it is 3½ inches.

In order to solve these problems of land measurement, the geodetic surveyors use a special way of measuring called *triangulation* (try·ang′you·lay′shun). This name comes from the word triangle. A *triangle* is a closed figure with three straight sides. The corners are called *angles*.

Several thousand years ago, mathematicians began to study triangles. They discovered many interesting things about them. For example, they found that if they knew the length of one side of a triangle and the sizes of two of the angles, they could figure out the lengths of the other two sides without measuring them. Modern geodetic surveyors

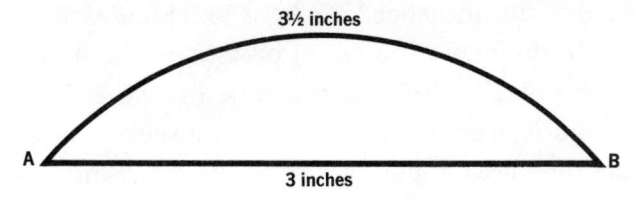

3½ inches

A   B
3 inches

UNDERSTANDING MAPS

use this fact when they want to find the straight distance between two points on land in spite of hills, valleys, and the curvature of the earth.

In triangulation, a field engineer and his surveying team work together. A triangular section of land is chosen, and a flag or marker is placed at one corner of the triangle. The distance between the other two corners is measured very carefully. This distance is the primary base line.

When a tape was used to measure the primary base line, it was important that the measurement be made over a fairly flat surface. Usually a long straight piece of road was chosen. The distance was measured at least thirty-two times, and the measurements were made at different hours of the day to be sure they were correct. Generally, the error in these measurements was not greater than one part in one million. This means that in measuring a piece of road five miles in length, the error usually was less than $\frac{1}{3}$ inch.

Until recently, most primary base lines were measured with an invar tape. Today a *geodimeter* (jee·odd′eh·meh·ter) is used. This instrument flashes light from one end of the base line to a reflector at the other. The reflector sends the light back to the geodimeter, which measures the time it took the light to travel down the base line and back to the instrument. If you know the speed of light, you can find the length of the base line. The round number most people use for the speed of light is 186,000 miles per second. In the measurement of base lines, the more accurate figure of 299,792.5 kilometers, or 186,282.4 miles, per second is used. If you multiply the speed of light and half the time it took the light to travel down the base line and back to the geodimeter, you will get an accurate measurement of the length of the base line.

Another instrument used to measure primary base lines is the *micro chain*. It uses radio waves which travel at the same speed as light. The results of using either the geodimeter or the micro chain are equal in accuracy to the measurements made with the invar tape. The advantages of using the geodimeter or the micro chain are that the measurements can be made more easily and more quickly, and they can be made over rough ground.

Although the accuracy of these distance-measuring instruments is amazing, the map-makers are still trying to improve them. *Lasers,* devices which can send a narrow, powerful beam of light as far as the moon, are now being added to the geodimeters. These pinpoint beams provide even more accurate measurements.

Once the length of the base line has been determined, the angles of the triangle at either end of the base line are measured with a *theodolite* (the·ah'do·lite). This contains a telescope and an angle-measuring instrument, and because levelness is so important, it also has a very good leveling device. A surveyor standing at one end of the base line measures the angle formed between the other end of the base line and the flag or marker. A similar measurement is then made with a theodolite at the opposite end of the base line. When the surveyor knows the length of one side and the sizes of two of the angles, mathematics can be used to find the lengths of the other two sides. Using either of these sides as the new base, the surveyor makes a new triangle and continues across the country.

Triangulation is often done at night. A powerful light

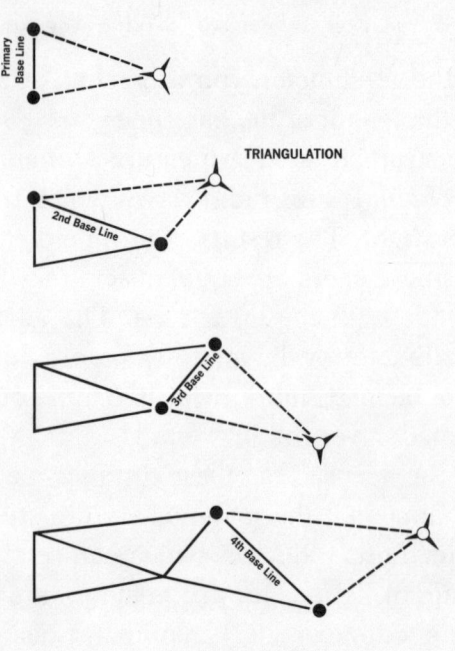

Triangulation

which can readily be seen through the theodolite replaces the flag or marker used for daylight measurements. In rough or wooded country, surveyors sometimes use metal towers like the tower shown on page 41. Such towers can be taken apart so that they can be moved easily, and put together again where needed. The towers can be extended to different heights, usually between 30 and 120 feet. They can also be set up in very rough country.

Once a fully extended tower was mounted in the bed of a rushing mountain stream. A surveyor who was new in the field made the mistake of looking down from the top of the tower. The rushing stream below made it seem as though the tower was falling, and the surveyor panicked and jumped.

When they use high towers, the surveyors are able to see

Assembling a Bilby triangulation tower

UNDERSTANDING MAPS

each other's lights over the tops of houses, trees, and even low hills. Sometimes, when the obstacles are too high, the surveyors use a helicopter which can hover directly over the spot chosen as a corner of a triangle. It is easy for the surveyor to see the hovering helicopter and so make his angle measurements.

Lights on mountain tops can be seen for a greater distance than lights on flat land, even when 120-foot towers are used. For this reason, the sides of a triangle usually measure about 10 miles on flat land and about 50 miles in the mountains. In cities, the sides of the triangle may be no greater than 2 miles.

Triangulation is also used to find the *elevation* (which means height) as well as the size of the land. It would be fun if you could measure the height of a mountain by triangulation. Of course, you don't have a Mt. Everest in your backyard, and even if you have a lesser mountain near your home, it would not be easy to do it without a theodolite. However, you can find the height of a tree, a telephone pole, or your house. It would be very difficult to climb up the side and measure it inch by inch, but you can do it by using a form of triangulation. With the help of the table on page 44, you will be able to find the height of objects which are too tall for you to measure directly.

NOW TRY THIS

Let's suppose you decide to measure a tree. The materials you will need are a tape measure, a yardstick, two stakes, a very long piece of string, and a friend to work with you. Tie one end of the string to a stake. Measure off

TREE EXPERIMENT

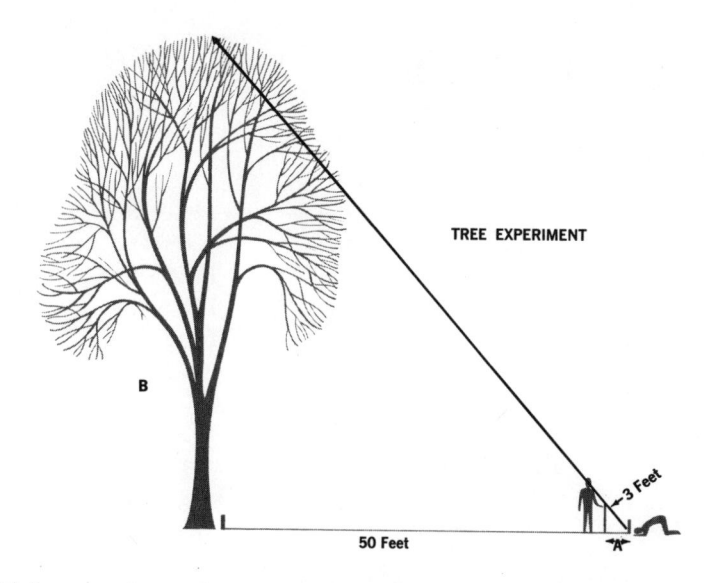

50 feet on the string and then tie it to the other stake. When the string is stretched out straight, the distance between the stakes must be exactly 50 feet. Put one stake in the ground at the foot of the tree. Place the other stake straight out from the tree, making sure the string is tight. Kneeling at the second stake and with your eye as close to the ground as possible, sight the top of the tree. Now ask your friend to hold the yardstick straight up and down and move it across the ground along the line of the string.

When the top of the yardstick and the top of the tree are in line, tell him to mark the spot. Then measure the distance from the stake where you were kneeling to the spot he has just marked. In the table on the next page, we will call this distance A. Let us suppose the distance A is 29 inches. Look in column A in the table until you find 29 inches. Now find the number on the same line in column B. It is 62 feet. Sixty-two feet is the height of the tree.

TABLE FOR FINDING HEIGHTS

A (Ground distance)	B (Height)
12 inches (1 foot)	150 feet
13 inches	138 ½ feet
14 inches	128 ½ feet
15 inches	120 feet
16 inches	112 ½ feet
17 inches	106 feet
18 inches (1 ½ feet)	100 feet
19 inches	94 ¾ feet
20 inches	90 feet
21 inches	85 ¾ feet
22 inches	81 ¾ feet
23 inches	78 ¼ feet
24 inches (2 feet)	75 feet
25 inches	72 feet
26 inches	69 ¼ feet
27 inches	66 ⅔ feet
28 inches	64 ¼ feet
29 inches	62 feet
30 inches (2 ½ feet)	60 feet
33 inches	54 ½ feet
36 inches (3 feet)	50 feet
39 inches	46 ¼ feet
42 inches (3 ½ feet)	43 feet
45 inches	40 feet
48 inches (4 feet)	37 ½ feet
4 ½ feet	33 ⅓ feet
5 feet	30 feet
5 ½ feet	27 ¼ feet
6 feet	25 feet
6 ½ feet	23 feet
7 feet	21 ½ feet
7 ½ feet	20 feet

(*The values in column B of this table are approximate. They were obtained from the solution of similar triangles.*)

MAPPING THE LAND: MAKING THE MEASUREMENTS

This is a rough way of measuring and, although the result may not be accurate, it is close. The slope of the ground and the straightness of the tree and the yardstick are just some of the things which might cause errors. But you do have a fair estimate of the height of the tree. You and your friend are like the surveyors in the field. You have found the measurements. Just as the men in the office do the mathematics for the surveyors, the table does the mathematics for you. Also like the surveyors, you have been working with triangles. Look at the picture on page 43 and you will see that you have used two triangles in measuring.

The surveyor measures the distances between key points and also their heights, or elevations, as he crosses mountains and rivers and moves through cities and jungles, but he does not draw a map. His findings are written in a log or field record and sent back to the survey office.

Another group of field men are the photographers and airplane pilots. As early as 1879, field men carried cameras as well as theodolites when they scaled high mountain peaks. With these cameras, they could take a series of pictures in all directions. This process was especially useful in producing some of the early maps of Alaska.

Cameras also were attached to kites and balloons. During the Civil War, cameras on balloons were used to map the positions of enemy troops. Photographs from balloon- and kite-mounted cameras were interesting, but not very useful for civilian map-makers, since the exact positions of the swaying kites and balloons could not be controlled. The map-makers had to wait for the invention of the airplane to obtain aerial pictures which were really useful.

The first big *aerial* project began in 1924, when the U.S. Coast and Geodetic Survey used photographs taken from airplanes to map the Mississippi River Delta. The following year, the Hamilton Rice Expedition was able to map, from the air, the impassable rain forests around the Rio Negro in Brazil. Today, photographs of the land are commonly used for map-making everywhere in the United States and the rest of the world. Whether the pictures are taken from a mountain top or from an airplane, the use of aerial photographs in making maps is called *photogrammetry* (fo'toe·gram'eh·tree).

A photographic flight must be planned with great care. The area to be mapped is divided into long narrow sections called *strips*. Each strip is planned so that it will overlap the strip on either side. The pilot must be very skillful, for the plane must fly on a straight line down each strip.

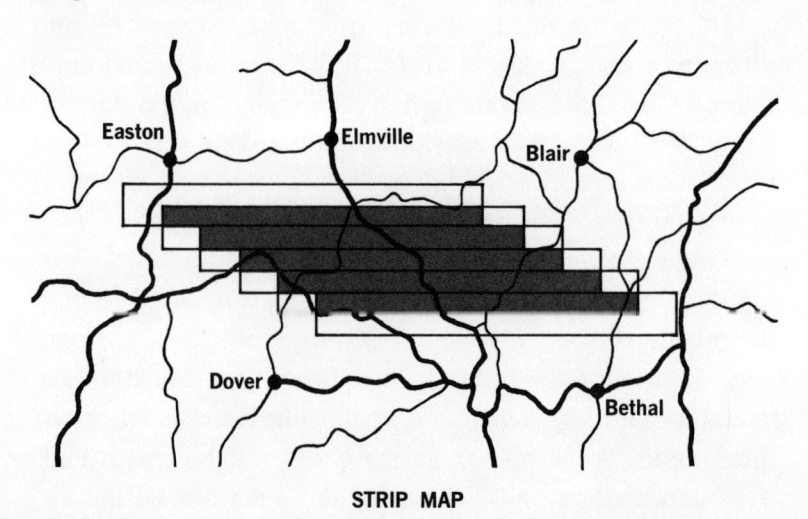

STRIP MAP

In planning a flight, the map-maker must consider the time of day, the season, the air currents, and the weather. The clearest pictures can be made in the spring or fall, when there are neither leaves on the trees nor snow on the ground. The best pictures are taken between 10 A.M. and 2 P.M., when the shadows are shortest. The pilot needs to know all about the air currents above the area. While it might seem that more detailed pictures could be made by flying low, strong currents and air pockets near the ground make it impossible to keep the plane steady and exactly on course. Moreover, many more flights would be needed to photograph the same area. In the eastern part of the country and in most densely settled areas, flights are planned at 1,350 feet. In mountainous sections and in other sparsely populated places, the photographs are taken from greater heights and so cover larger areas.

Most important of all, the air must be cloudless and clear. Ideal photographic days are very rare in some places. When the Brooks Range in Alaska was being photographed by the map-makers, only five days occurred in the entire summer that were suitable for photography. It is not unusual for photographic planes to start working somewhere in the western part of the country during a period of good weather and follow that clear weather as it moves eastward across the continent, taking photos wherever new map-making projects are under way.

Most planes used for aerial photography are privately owned and operated. The planes are equipped with two cameras which simultaneously take pictures of the same area. A tremendous roll of film is used in each camera. The finished negatives are 9 inches by 7 inches. When the pictures are taken at a height of 1,350 feet, 15 negatives are needed to cover the area shown on most quadrangle maps today. Each inch of the negatives shows about ½ mile of land. When the photographs are taken at a height of 1 mile, only 6 negatives would cover the same area, and each inch on the negative would show about 2 miles of land.

The aerial photographer is like the surveyor; he gathers the information. The map-makers, working in an office that is often many hundreds of miles away from the area, change the surveyor's measurements and the photographer's pictures into our familiar maps.

Mapping the Land: Drawing the Maps
4

During World War II, the military forces needed an accurate map of Tibet. This little known Asian country, on whose southern border Mt. Everest rises, had never been mapped scientifically.

A surveyor was instructed to pick a key spot in Tibet and determine its exact location and altitude. The surveyor packed his instruments and traveled as far as possible by jeep. Then he continued on foot for several weeks, crossing the cold, snow-covered mountainous country. Finally, the surveyor reached a small village in a valley where he made the necessary observations and measurements, and then he took a series of pictures of the village from the surrounding mountains. These pictures were to be used to help identify the village on the films made by photographic flights. His work completed, the surveyor hiked back with his instruments, films, and log.

Meanwhile, men in airplanes were photographing the land by making dangerous flights back and forth above the bleak Himalaya country. The films and log of the surveyor and the films from the flight cameras were rushed to the *cartographers* (car·tah′gruff·ers), the men who draw the maps.

UNDERSTANDING MAPS

Unfortunately for the project, small Tibetan villages are all built on the same pattern. The cartographers were unable to recognize the village where the measurements were made in spite of the surveyor's pictures. Even the surveyor, who was asked to help, could not identify the village. Without this information, the flight photographs were useless. So it was necessary for the surveyor to make the long trip back into the mountains to find the village. This time, he placed a large marker on the ground so that it could be seen clearly in the next set of flight pictures. With this one village identified on the pictures, the cartographers were able to use the flight film to produce a reasonably accurate map of this previously unmapped land.

In mapping the United States, cartographers have many readily recognizable points where ground crews have made careful observations and measurements. In 1819, the United States government began establishing such key points. The first measurements were made near Gravesend Village, Long Island, and near English Creek, New Jersey.

The original plan was to survey the east coast of the young nation. From this *Coast Survey,* as it was called, a complete network or grid of key locations across the entire country and all of its island possessions has been developed. The establishment of this grid and the responsibility for its accuracy are part of the work of the United States Coast and Geodetic Survey. At each of the key locations, the bronze bench mark of the Coast and Geodetic Survey can be seen.

Before photogrammetry was developed, cartographers had much more freedom in drawing maps. They often accompanied the surveyors and drew the maps as the measurements were being made. They showed the contours by using many short lines to give the impression of lights and shadows on a hill. Using lines this way to show contours is called *hachuring* (ha·shur′ing). The hachuring lines on a map are drawn in the direction the water flows across the surface of the land. Because hachuring requires such great skill, a simpler form, the *caterpillar map,* was devised.

If the results of the hachured maps were less accurate than those produced by modern methods, the finished maps often were works of art. In fact, many of the early cartographers were artists who etched their drawings on the copper plates from which the maps were printed. James McNeill Whistler, who is most widely known for his portrait of his mother, began his career as a cartographer.

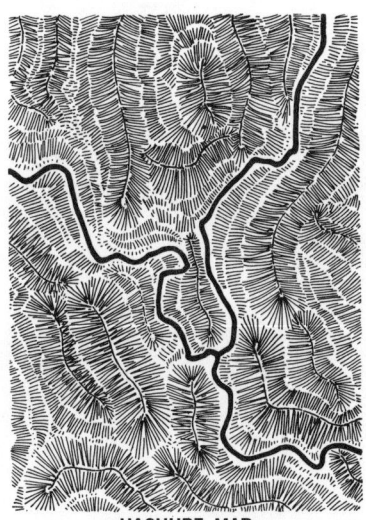

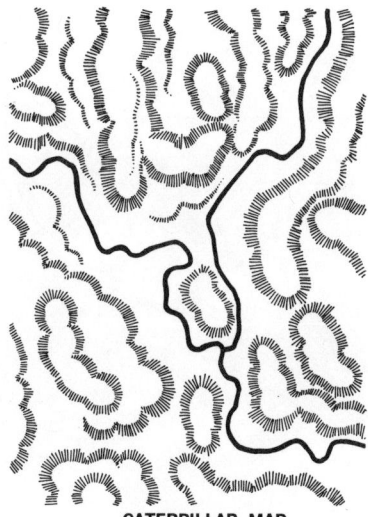

HACHURE MAP   **CATERPILLAR MAP**

UNDERSTANDING MAPS

Most map-makers today use *contour lines* to show the variations in height on the surface of the earth. Contour lines are curved lines which connect areas which have the same elevation. For example, all the places on a hill which are 100 feet above sea level would be joined by one contour line. All those 150 feet high would be joined by another contour line. Sometimes the elevations are printed along the contour lines.

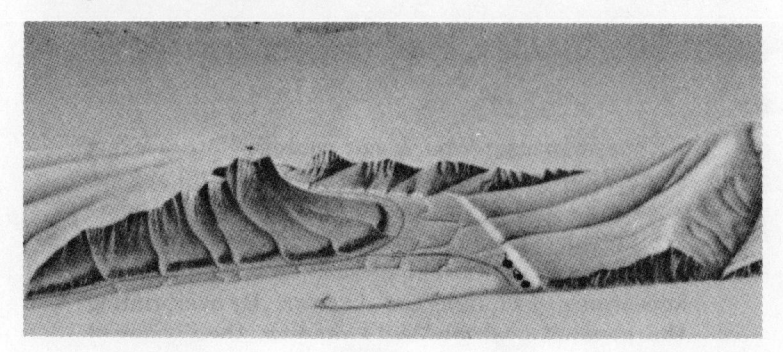

HILL

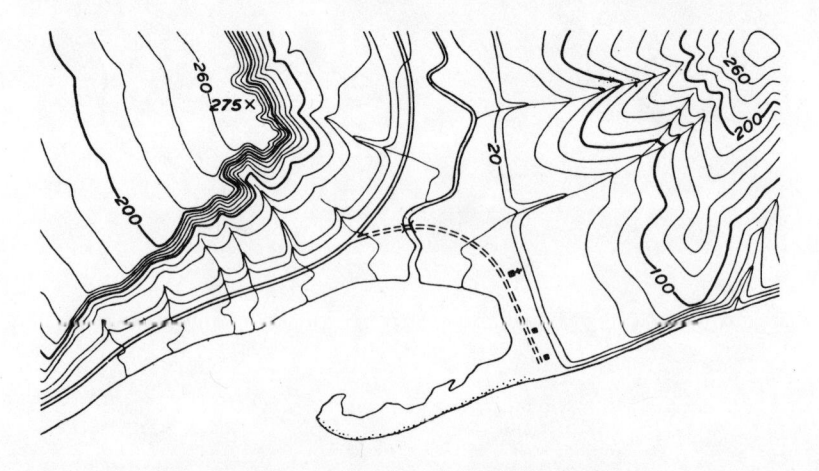

CONTOUR LINES OF THE HILL

MAPPING THE LAND: DRAWING THE MAPS

The simplest maps to make from photographs are *mosaic maps.* These photomaps are prepared by fitting together individual photographs to form a view of the entire area. Grids, names, symbols, and scale are added to complete the map. Such maps are not as accurate as line maps, but they give a clear view of an entire area, and they can be prepared more quickly and less expensively. They often are used in the planning stages of civil works such as the construction of roads, dams, or airfields.

The modern cartographer uses many complex instruments to help him prepare line maps. His first task is to select the grid on which the map will be placed and to draw the grid lines on a plastic sheet which will become the *base map.* Then the cartographer locates each of the control points, such as the bench marks. Next he draws in such features as rivers, roads, and the contours of the land. To do this, he uses a *stereoscopic* (stehr'ee·oh·skahp'ic) *plotter* so that he can view the films from both cameras at the same time. This produces a three-dimensional image of the land, an image which has depth as well as length and width.

The two cameras in the airplane were placed at a fixed distance from each other and each took a picture of the same area at precisely the same time. These are stereo pictures. You can find out why it is important to use stereo pictures.

NOW TRY THIS

Hold a pencil at arm's length in a line with your nose. Close one eye. Move your head and arm until the pencil appears to be lined up with some object on the wall in front of you. If you do not know about how far the pencil is from the wall, can you guess the distance? Without moving, close the other eye and open the first eye. Where does the pencil appear to be now in relation to the object on the wall? Can you judge the distance between the pencil and the wall? Open both eyes. Can you judge the distance now?

Your eyes, like the two cameras in the airplane, are a fixed distance apart. Each eye sees the pencil from a slightly different position. Your brain merges the two images to produce a stereo image. A stereoscopic plotter uses two projectors to merge rays of light from the two films and produces a stereo image. One projector uses red light and the other uses blue light. The cartographer wears glasses with one red and one blue lens when he views the image.

A dot of white light is projected through a plate onto the colored image. The cartographer moves the plate so that the dot always rests on the surface of the image as it follows whatever feature he wishes to include on the line map. He can move the dot along the course of a river, following it to the smallest feeder streams. He can follow highways or outline airfields. As the dot moves, a needle on another part of the instrument reproduces the movement on the plastic base map. The needle scratches out the course of the river or whatever else the cartographer is following. In this way, the cartographer traces the position and contour of all the features he wishes to appear on the finished map. The cartographer can detect differences of

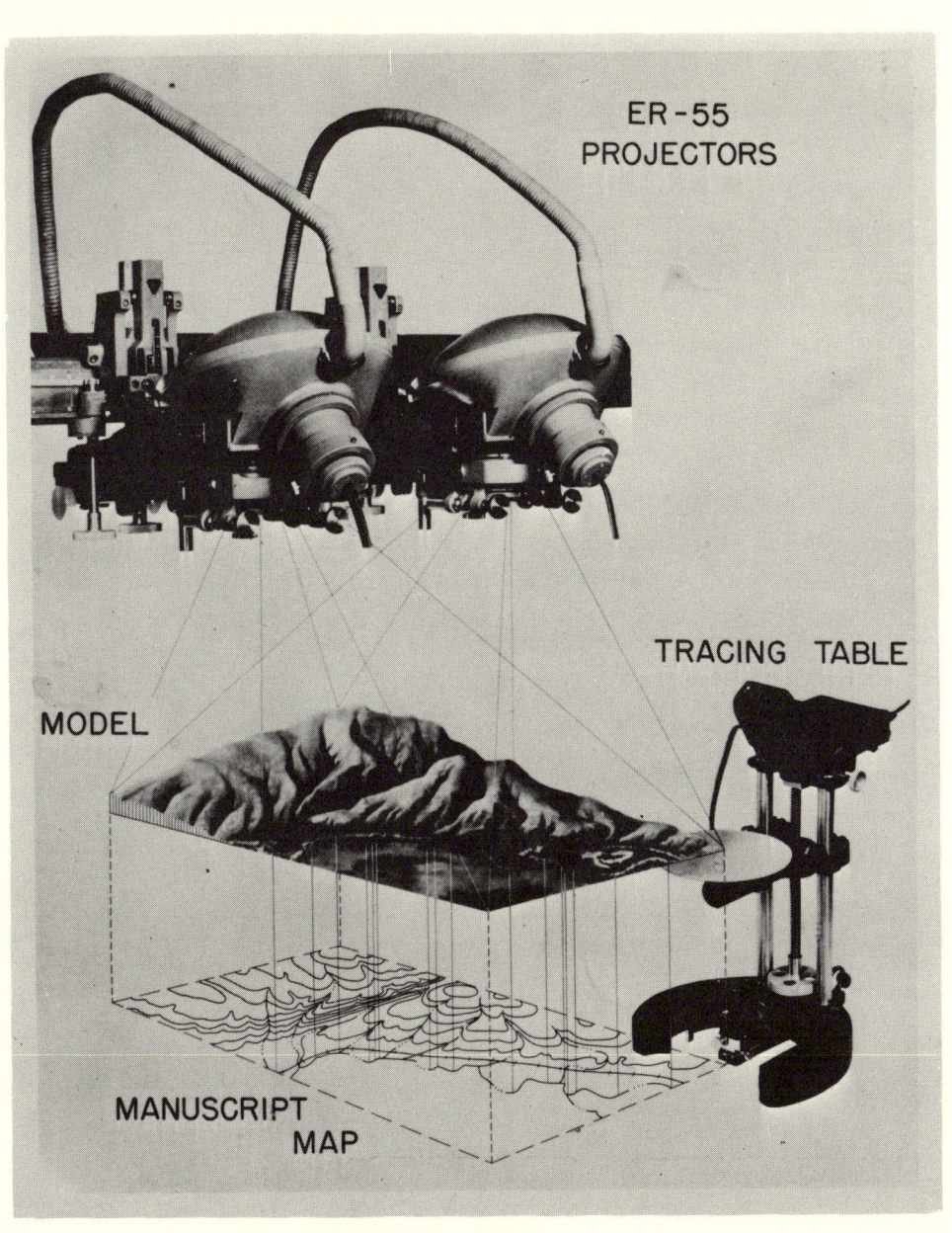

Stereoscopic plotter

only a few feet in elevation and they can be plotted easily.

The cartographer who uses a stereoscopic plotter must have good stereo vision. All people who can see out of both eyes have stereo vision, but not everyone can use a plotter. Special tests are given to people who plan to work with this instrument.

Copies of the base map are sent to the men in the field who resurvey selected areas of the map to be sure that no mistake was made during the stereoscopic plotting. Some field men who have returned to make these checks have heard complaints from local people. These citizens assume that the field men were careless and did not do the survey correctly the first time. But government mapmakers know that their maps will be used for many pur-

Checking in the field

poses where accuracy is important. It is worth the time and money to field-check the base map at this point.

The field men also check the names of all landmarks shown on the map. Sometimes they find that some landmark, such as a small creek, has more than one name, or that there are several ways of spelling its name. Then it is their job to find out as much as possible about each name or spelling. They find out which name is used most often by the people who live there, and they try to learn the history behind each of the names. All the information they can gather is sent to the Board of Geographic Names. This government agency reviews the reports from the men in the field and assigns official names to the landmarks. Every four months, a list of official names is published.

When the base map has been checked and all corrections have been made on the plastic sheet, the work of the *scribers* begins. Theirs is a careful, painstaking job. The plastic sheet is placed on a table which has a glass top and a light beneath it. A negative is placed on top of the base map. This negative is made of a special plastic called *mylar*. Mylar is tough and does not stretch or change shape because of changes in temperature. This is very important because several mylar sheets are made for each map. When the sheets are placed on top of each other, each mark made on the sheets by the scribers must be in exactly the correct place. The mylar sheet to be engraved has a smooth coat of bright yellow on which very thin scratches can be made easily. The scriber, who is also called a negative engraver, uses a pen-like instrument to mark on the yellow. The point of the instrument is a steel phonograph needle which has been sharpened to an exact width. To

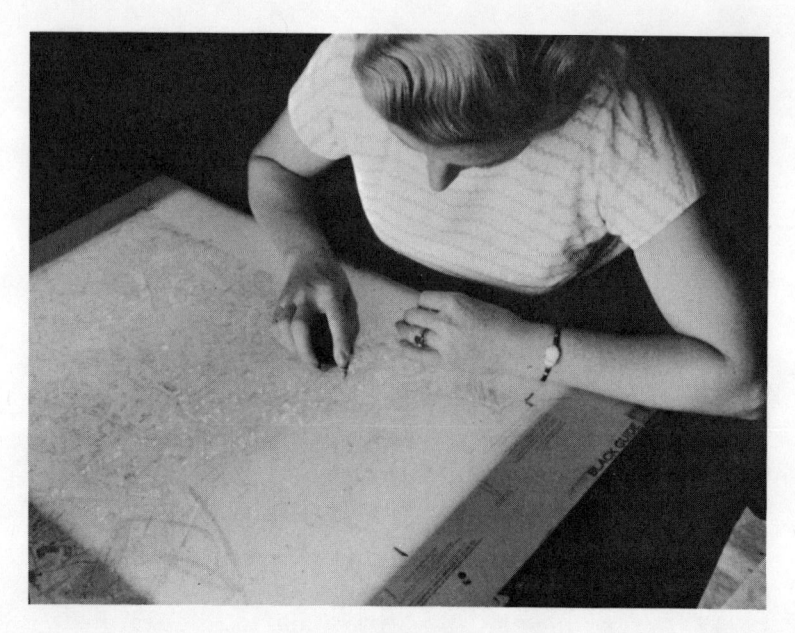

Scribing

draw roads, a needle is sharpened to such a fine point that it draws a line only $2/100$ centimeter in width. Some contour lines are made with a needle that makes a scratch about $1/100$th as thin.

The scriber copies each line on the map in the required width. If anything on the map does not appear quite right, he stops, and it is checked against the original field report or photograph. A good scriber engraves an average of about 8 square inches of a quadrangle map a day.

The *color separators* are the next people to work on the map. Using the base map, the separator decides where each color is to appear on the map. Four colors are used on quadrangle maps, and a separate plastic sheet must be made for each color. This is done by peeling off the yellow coat from a sheet wherever the color is to appear on the finished map.

On still another sheet, all the symbols to be used are placed in their exact locations. More than 500 symbols may appear on a single quadrangle map. An editor decides what size and kind of type must be used for every printed word on the map and its margin. These words may include the names of as many as 3,000 places. They are printed on clear adhesive tape, which is then put on still another sheet.

After a final check for accuracy, the sheets are sent to the printing office where the maps are printed. At last, they are ready to be sent wherever they are needed. To complete a modern quadrangle map takes about three years from the time planning begins until the map is ready for distribution, and three workers are needed in the cartography department for every man in the field.

The Army Corps of Engineers is now using new machines which plot maps electronically. In this system, called UNAMACE, meaning Universal Automatic Map Compilation Equipment, computers replace the stereoscopic plotters. Such automation can plot a map in 24 hours that formerly took at least 6 months to complete. However, even with UNAMACE, some handwork is necessary, and field checks still have to be made.

The men who map the land have had to face many problems, but they always have been able to work out answers. If mountains were in the way, they climbed to the top and went on with the triangulation. If the tropical forest was too thick, or the Antarctic snow too deep, they used aerial photography. When helicopters were available, they were used as triangulation stations or as substitutes for pack mules and jeeps. New instruments that use light and radio

UNDERSTANDING MAPS

waves to measure distances replaced the chains and tapes.

Each of these methods of land measurement depends in part on landmarks. But only 28 per cent of the earth is land, and the surveyors cannot set up their theodolites on the pitching waves. Even aerial photographs cannot help because a picture of the North Pacific might look exactly the same as one of the South Atlantic. There are no visible landmarks to help, and yet map-makers can and do chart the vast expanse of rolling ocean.

Mapping the Sea: Finding the Latitude
5

The ice caps of the polar region, which send blasts of cold weather out over the warmer parts of the earth, also create nightmares for the men who sail the ships. With their wonderful charts, today's sailors can check quickly the coast of the mainland or the location of any island. But far to the north and the south, the polar ice caps feed into the sea temporary islands which are always in motion. These islands of ice, called *icebergs,* are found in many shapes and sizes.

Most of the icebergs which originate in the Antarctic are flat or have slightly rounded tops. They may be as large as the 2,000 square miles which broke from the Ross Ice Shelf, or small enough to fit into a refrigerator. Small icebergs are called *growlers*. But giants or growlers, the icebergs from the Antarctic menace the ships that go around Cape Horn.

The icebergs that break off the glaciers of Greenland look like fairy castles crowned with spires. With only $\frac{1}{9}$ of their bulk showing above the water and a deadly $\frac{8}{9}$ hidden beneath the surface, sprawling in unknown shapes, the icebergs sail majestically down the Labrador Current straight into the main shipping lanes that connect Europe and America.

For many years only the eyes of the lookouts in the crow's-nests, peering into the dreary fog or inky night, protected the ships. Many were wrecked by the silent floating islands. Perhaps the most famous was the *Titanic,* which everyone believed was unsinkable. This great passenger liner, built at a cost of many millions of dollars, sailed from England on April 10, 1912, on her maiden voyage to America. Four days later she was only a wrecked hull lying at the bottom of the North Atlantic, with 1,517 of her 2,224 people lost. An uncharted island of ice had sunk the "unsinkable" ship before she could complete her first voyage.

Lighthouses, beacons, and maps of the area can protect the sailor from the dangers of a rocky shore. Even though icebergs do not come equipped with lighthouses or radio-signaling devices, their positions and probable courses can be mapped. The tragedy of the *Titanic* made people all over the world aware of the danger of icebergs, and as a result, in 1912, the International Ice Patrol was formed. Today its boats and planes prowl the northern seas in search of icebergs. From the information they supply, the U.S. Hydrographic Office issues a monthly *pilot chart* showing the temperature of ocean water and the ice fields located during the previous month. Since these move about so freely, the Hydrographic Office sends out special bulletins to ships at sea whenever necessary.

But the captain of the *Titanic* was not thinking about future maps on that foggy April night. What he needed was help, and he needed it quickly. If you need help at

MELVILLE BAY

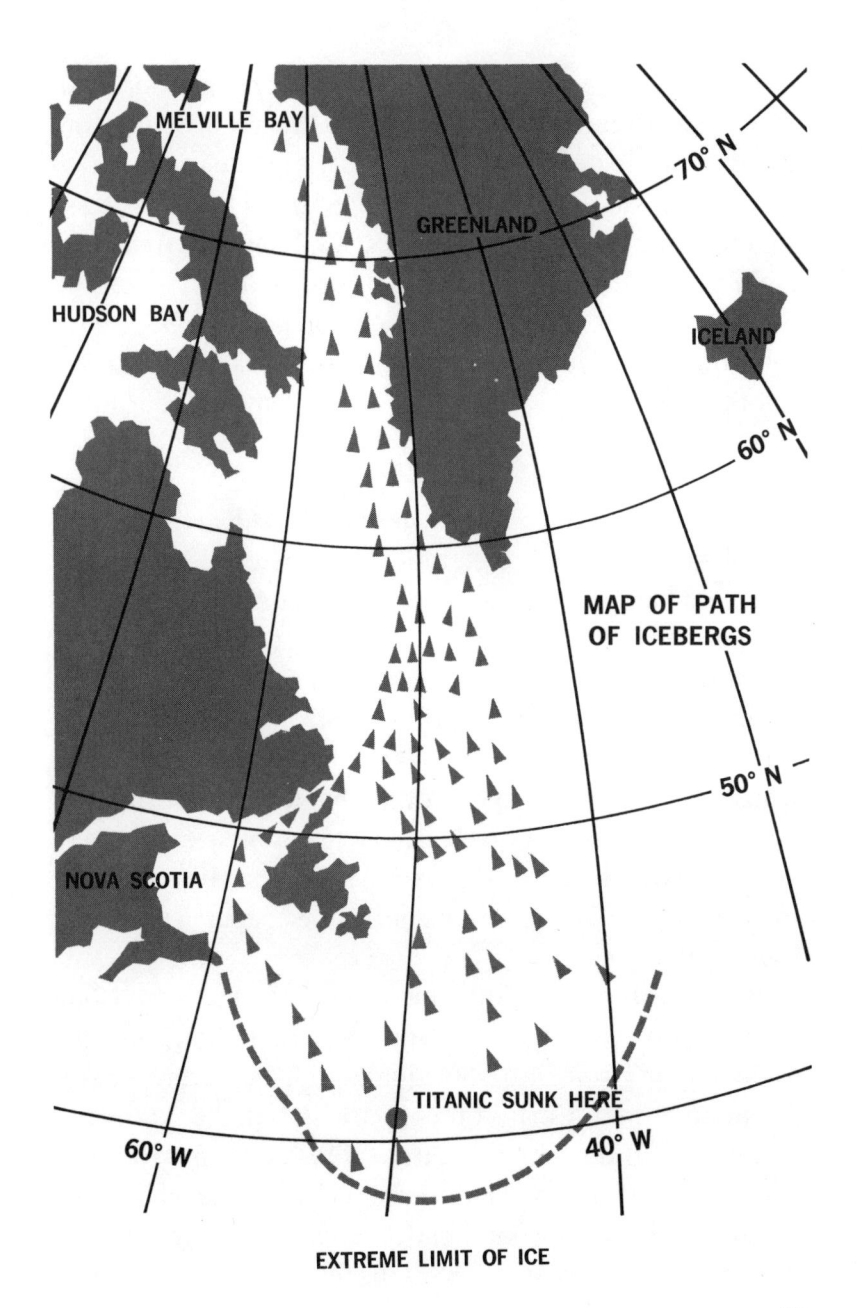

GREENLAND

HUDSON BAY

ICELAND

70° N

60° N

MAP OF PATH
OF ICEBERGS

50° N

NOVA SCOTIA

TITANIC SUNK HERE

60° W   40° W

EXTREME LIMIT OF ICE

home, you can call the police or fire department and tell them your street and house number. A ship at sea has no street or house number, or even a city or state name which can identify its location. But each ship has an "address" which changes as the ship moves. And so the wireless on the *Titanic,* sending through the air for the first time in history the new international call for help, "SOS," gave this address: 41°45′ N. Lat.; 50°14′ W. Long. With this information, the steamer *Carpathia* traveled 58 miles across the trackless ocean directly to the wrecked *Titanic.*

When you wish to measure the distance between two points, you use *linear measurement.* It deals with distances such as inches, feet, miles, and meters. Linear measurement can be used on either flat or curved surfaces.

There is another way of measuring. It is called *angular measurement,* and it shows the *ratio* or relation of a part of a circle to the whole circle. If you fold a piece of paper exactly in the middle and then tear it, each piece is ½ of the whole sheet. If you fold one of the halves and tear it, each new piece is ¼ of the whole. If you tear a quarter, you will get pieces which are ⅛ of the whole. This can be continued through ¹⁄₁₆, ¹⁄₃₂, ¹⁄₆₄, and until the paper is smaller than confetti.

It is possible to divide a circle in the same way. Long ago the people of ancient Babylon (baa′bih·lon) decided to divide the circle into 360 equal parts, and today we still use their system. Each of these parts is called a *degree,* so there are 360 degrees in a circle. This is often written as 360°. The mark ° means degree. Each degree can be divided into 60 equal parts called *minutes,* written as 60′. And, if one wishes an even smaller measure, each minute can be divided into 60 equal parts called *seconds,* written as 60″. For navigation purposes, generally only degrees and minutes are used.

Man always has been very curious about this world in which he lives. The people of ancient Greece were unable to explore much of the surface of the earth because they had only very primitive means of travel. They used their minds instead to find out things which their eyes could never see. They invented *geometry* (gee·om′eh·tree) to help them measure the earth. In fact, the word geometry comes from two Greek words which mean to measure the earth.

All measurements must start somewhere. Any flat area, such as a field, has a beginning and an end, so it is always possible to find an edge from which to start measuring. But look at a ball. There is no beginning and no end, and our earth is shaped roughly like a ball.

To measure the earth, the people of ancient times chose two imaginary lines. One is a straight line running through the center of the earth from the geographic North Pole to the geographic South Pole. This is called the *earth's axis.*

The second imaginary line they chose for measuring is a curved line which goes around the earth and is at all places halfway between the North and South Poles. This line is called the *equator* because it is equally distant from both poles, and divides the earth exactly in half. The distance from every point on the equator to either pole is one-fourth of a circle going completely around the earth and passing through each pole. Since a circle is divided into 360 degrees, the distance from any point on the equator to either pole is one-fourth of 360 degrees or 90 degrees. The dis-

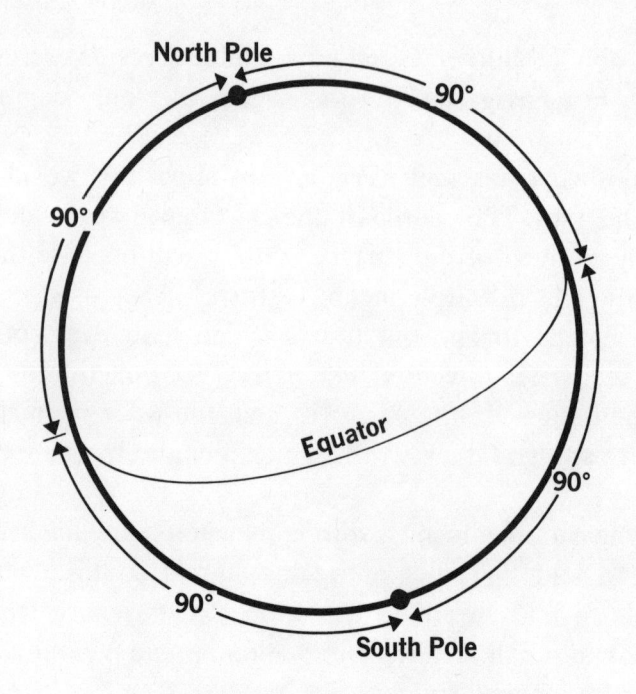

THE DISTANCE FROM THE EQUATOR TO EITHER POLE IS ¼ OF A CIRCLE OR 90°

tance in degrees north or south of the equator is called the *latitude*.

A position north of the equator is always written as so many degrees North Latitude, and south of the equator as so many degrees South Latitude. For example, Cocos Island, in the Pacific Ocean, is about 5 degrees north of the equator and is noted on sea charts as 5°30′ N. Lat. Easter Island is about 27 degrees south of the equator and is written as 27°3′ S. Lat.

When two or more lines are the same distance apart for their entire length, they are called parallel lines. On a globe you will find a number of circles which are parallel

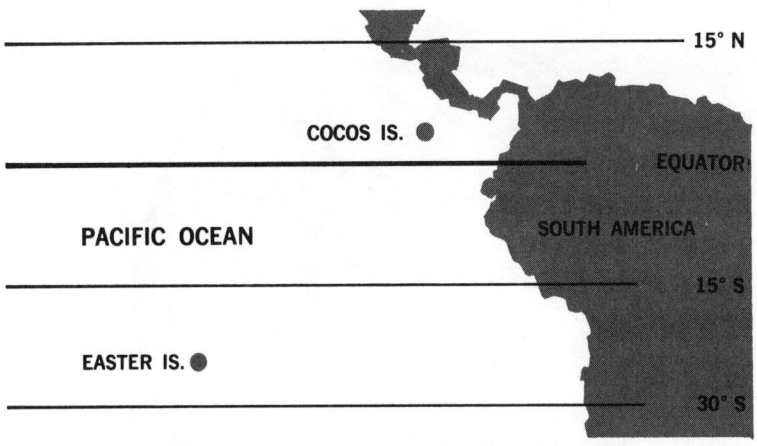

POSITION OF COCOS AND EASTER ISLANDS

to the equator. These are *parallels of latitude*. Notice that as you come closer to the poles, the distance around each parallel becomes shorter. A parallel is a guide to help locate a spot on the globe.

If you want to find Bear Island, which is at 75° N. Lat. in the Arctic Ocean, you do not need to go all the way back to the equator to start measuring. You can look for it between the 70th and 80th parallels.

A sailor at sea cannot reach out across the water to the imaginary line called the equator when he needs to measure the position of his ship. But just as he has learned to use the North Star to find direction, he can use it also to determine latitude. To understand the way in which the Pole Star helps a sailor, you must imagine that the part of the sky which can be seen from the Northern Hemisphere can be compared with the inside of a great bowl hanging over the earth with the Pole Star at its center. The North Pole is almost directly under the North Star. (Actually, it is about 1 degree off center.)

GLOBE SHOWING PARALLELS OF LATITUDE

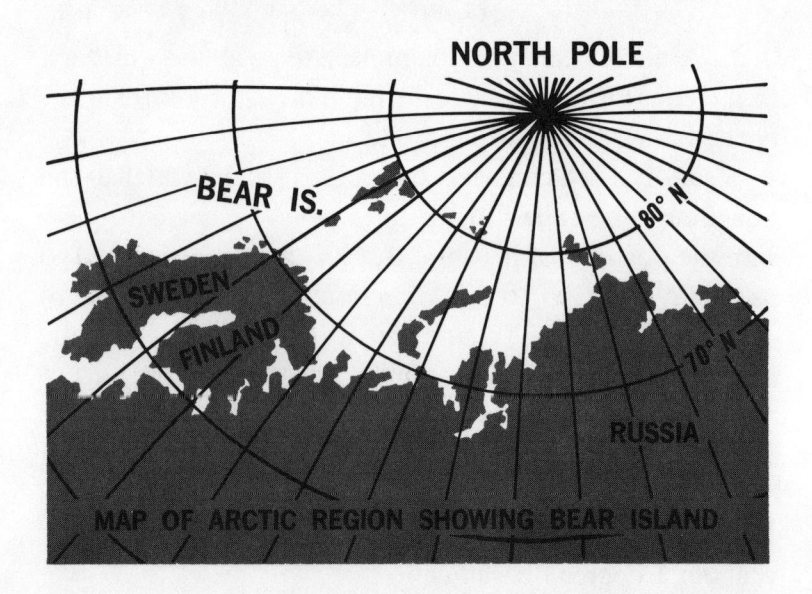

MAP OF ARCTIC REGION SHOWING BEAR ISLAND

As the earth spins around its axis, it appears as though the whole "bowl" of the sky is turning, with only the Pole Star standing still. If you hold a large salad bowl upside down over a globe of the earth, you will have a very rough model of the universe. Hold the center of the bowl directly over the pole and spin the globe slowly. It is easy to see that no matter where you stand, except at the North Pole itself, the Pole Star will always be north of you.

When a sailor at sea looks into the distance in any direction, the earth and sky seem to be touching. As he slowly turns around, it seems as if the "bowl" of the sky meets the earth in a huge circle with him at the center. The circular line where earth and sky appear to meet is called the *horizon*. On the open sea, you can see the horizon very easily. You also can see parts of the horizon at the beach or any place on land where there are no hills or buildings to block your view.

Even before the time of Christ, people working with geometry found a way to measure latitude. They discovered that the angle formed by two imaginary lines, one from the horizon to the eye, the other from the eye to the Pole Star, is equal to the *latitude*. Of course a sailor cannot draw these imaginary lines, but he can do what the hunter does when he aims a gun. When a hunter lines up a rabbit in the sights of his rifle, he is drawing an imaginary line from his eye to the animal. In the same way, a sailor can "shoot" both the Pole Star and the horizon. The size of the angle between these two lines is roughly his latitude.

The nearer the sailor is to the equator, the smaller the angle between the imaginary lines. As a ship sails north, the angle becomes larger. When a ship is off the coast of

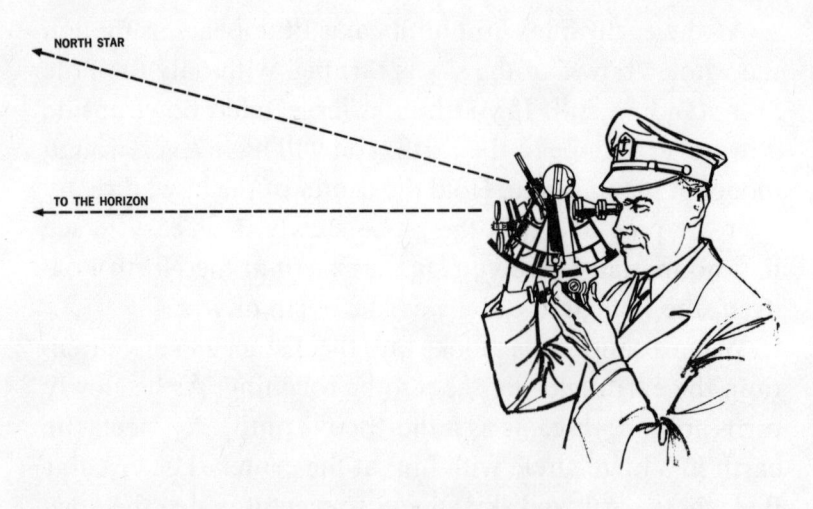

SHOOTING THE HORIZON AND NORTH STAR

St. Augustine, Florida, the angle is 30 degrees and the latitude is 30 degrees North. If the ship sails north to White Bay, Newfoundland, the angle will then be 50 degrees and the latitude will be 50 degrees North.

A sailor can measure this angle by using a *sextant*. This instrument which has two arms for sighting was invented in 1731. It uses a system of mirrors and lenses on the arms so that the sailor can shoot the North Star and the horizon at the same time. It also has an accurate scale to measure the angle between the arms.

Today the sailor can find his latitude more exactly by correcting such known errors as the fact that the Pole Star is about 1 degree off true north. He does this quickly and easily by using one of the tables in the *Nautical Almanac,* which is published each year by the Naval Observatory in Washington, D.C.

You can find the latitude of your house by shooting the

MAPPING THE SEA: FINDING THE LATITUDE

North Star. Your results may not be absolutely accurate for many reasons. For one thing, you probably will not have a sextant, but you can make a simple angle measuring instrument.

NOW TRY THIS

You will need a straight piece of wood about 2 feet long, two screw eyes, a piece of string, a stone or other weight, a piece of adhesive tape, a thumbtack, and a protractor. A *protractor,* which is a half circle divided into 180 parts, or degrees, along its curve, is used to measure angles. (Remember, a complete circle has 360 degrees.) You may have a protractor, or you can buy one at a store

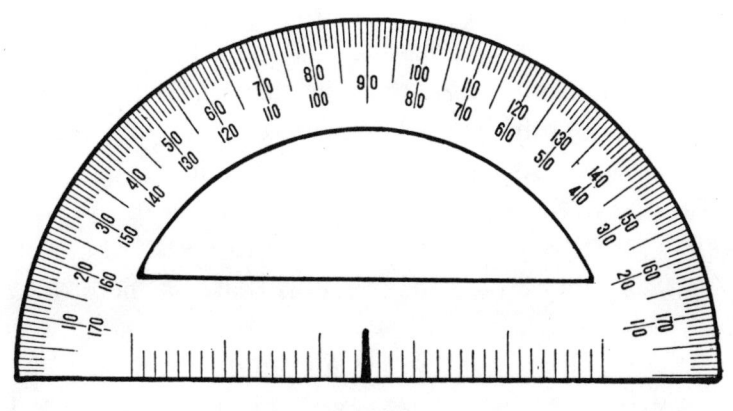

PROTRACTOR

71

that sells school supplies. If you want to make your own, trace the picture on very thin paper, paste it on cardboard, and cut around the outer line.

Fasten one screw eye into the top edge of the piece of wood about an inch from one end. Screw in the other about three-quarters of the way down the stick in a straight line with the first screw. These are your sights. Tape the protractor securely to the side of the wood. Be sure the straight edge of the protractor is parallel to the top edge of the wood. Tie one end of the string to the stone or weight. Attach the other end with a thumbtack to the wood directly above the arrow on the protractor. Now your instrument is ready for use. For best results, have a friend help you.

When the instrument is held level, the string will pass over the protractor at the 90-degree mark. In this position, the wood is pointing along the imaginary line from you to the horizon. Find the Pole Star in the night sky. (See page 24.) Now hold the end of the wood close to your eye and shoot the Pole Star by lining it up through the sights.

Be sure to keep your instrument steady in this position while your friend reads the degree mark where the weighted string crosses the circular scale on the protractor. (Since there are two rows of numbers, he should read the smaller number.) Subtract this number from 90°. This is the size of the angle between the horizon and the Pole Star. It will give you an estimate of your latitude. Take turns shooting the star several times to get the most accurate reading possible. Parallels of latitude are numbered on the sides of a map. Check your results with the latitude of your home town as shown on a map.

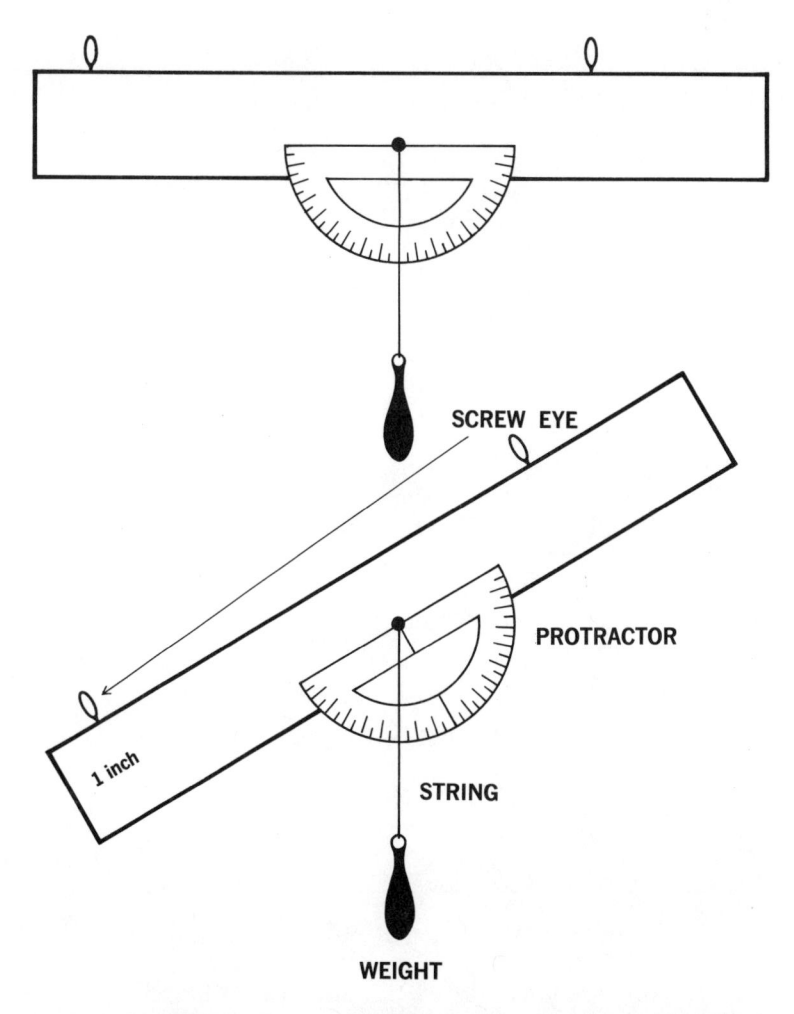

People in the Southern Hemisphere cannot find latitude by a pole star. Because of the curve of the earth, the Pole Star cannot be seen below the equator, and there is no star directly over the South Pole. However, the *Nautical Almanac* has tables which can be used to find latitude by shooting other stars, including the sun. Since it can be

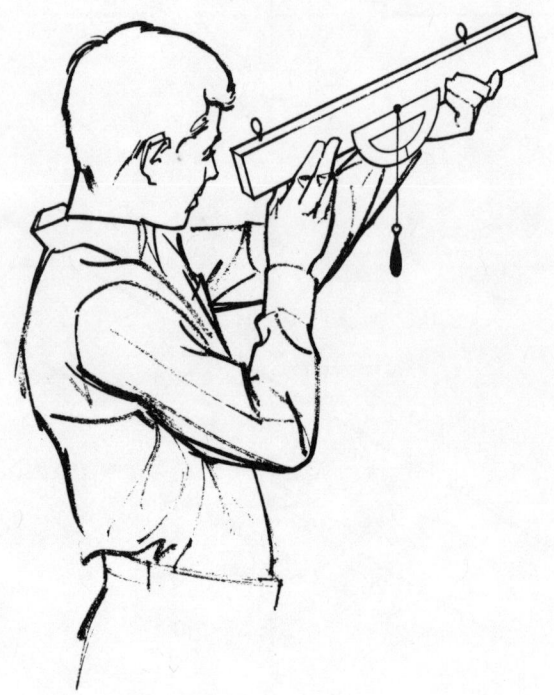

HOME MADE ANGLE-MEASURING INSTRUMENT

harmful to look directly at the sun for even a brief moment, never use your instrument to shoot the sun.

As part of an international project which began in 1900, five observatories were set up to study any variations of latitude. The observatories are all on the same parallel of latitude, 39°08′ N. Lat. Two of these stations are located in the United States and are operated by the

U.S. Coast and Geodetic Survey. One observatory is at Gaithersburg, Maryland, and the other is at Ukiah, California. The foreign observatories are on Sardinia, an island off the southwest coast of Italy, in Soviet Turkistan, which is in central Asia, and north of Tokyo on Honshu, the largest of the Japanese Islands. Each of the observatories uses the same type telescope to make nightly observations of 18 pairs of stars. Japanese scientists are in charge of collecting and interpreting the information gained from these observations.

For many years, it has been known that the axis of the earth wobbles slightly as the earth rotates. From the observations made at these observatories since 1900, the pattern of the wobble has been determined, and the scientists will now be able to report any changes. Moreover, as the information is collected over a longer period of time, it can be used to test a favorite theory of some scientists. These men believe that the continents are slowly drifting apart. Some scientists suggest that at one time Africa and South America formed one continuous land surface. If you look at a map which shows the east coast of South America and the west coast of Africa, you can see from their shapes that this is possible. The measurements being made by the variation-of-latitude observatories are so precise that they can be used with a computer to detect very small changes in the distance between two places. For example, it will be possible to note a continental drift as slight as one centimeter occurring over a period of many years.

Mapping the Sea: Finding the Longitude
6

A sailor who knows only a ship's latitude is in the same position you would be if you and a friend wanted to meet on the street so that you could go to the movies together, and if he told you, "Meet me on Main Street." You might find him, but what if Main Street is several miles long? And if he said, "Meet me on Maple Avenue," you might have the same problem. But if he told you to meet him at the corner of Main Street and Maple Avenue, you could find him easily. What you needed to know in order to pinpoint your meeting place was two sets of information.

In the same way, if you want to find an exact spot on a piece of paper, two measurements are needed, the distance from the top or bottom of the page and the distance from either of the sides. Suppose you want to start a drawing 1 inch from the top and 1½ inches from the left side. In the next diagram, the line *AB* is 1 inch from the top along its entire length. The line *CD* is 1½ inches from the left side for its entire length. No matter how long these lines are drawn, they will meet only at the point *X*.

One inch from the top of the page could be anywhere along the line *AB*. One and one-half inches from the left

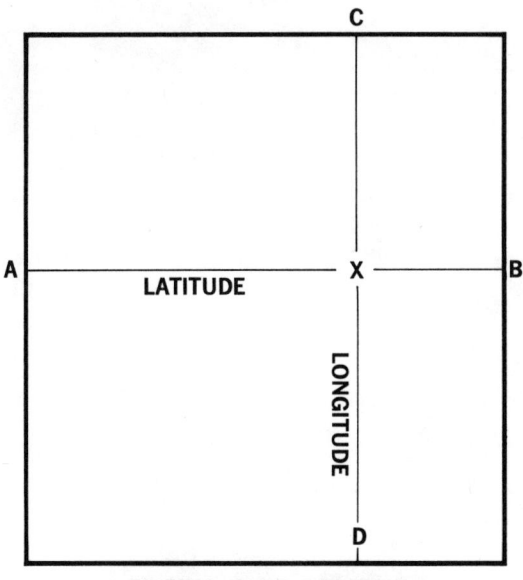

**FINDING YOUR POSITION
BY USE OF LATITUDE AND LONGITUDE**

could be anywhere along the line *CD*. But the point *X* where the lines cross, and only that point, is 1 inch from the top *and* 1½ inches from the left side.

A sailor who wants to give the location of his ship uses the same method. He gives two measurements. One is the distance north or south of the equator, the latitude. The other is the distance east or west of Greenwich, England, and is called the *longitude*. A line drawn through a given latitude and a line through a given longitude will cross at only *one* point on a map.

A line of longitude, or a *meridian* (muh·rid′ee·an) is a half circle running along the surface of the earth from the North Pole to the South Pole. In order to measure dis-

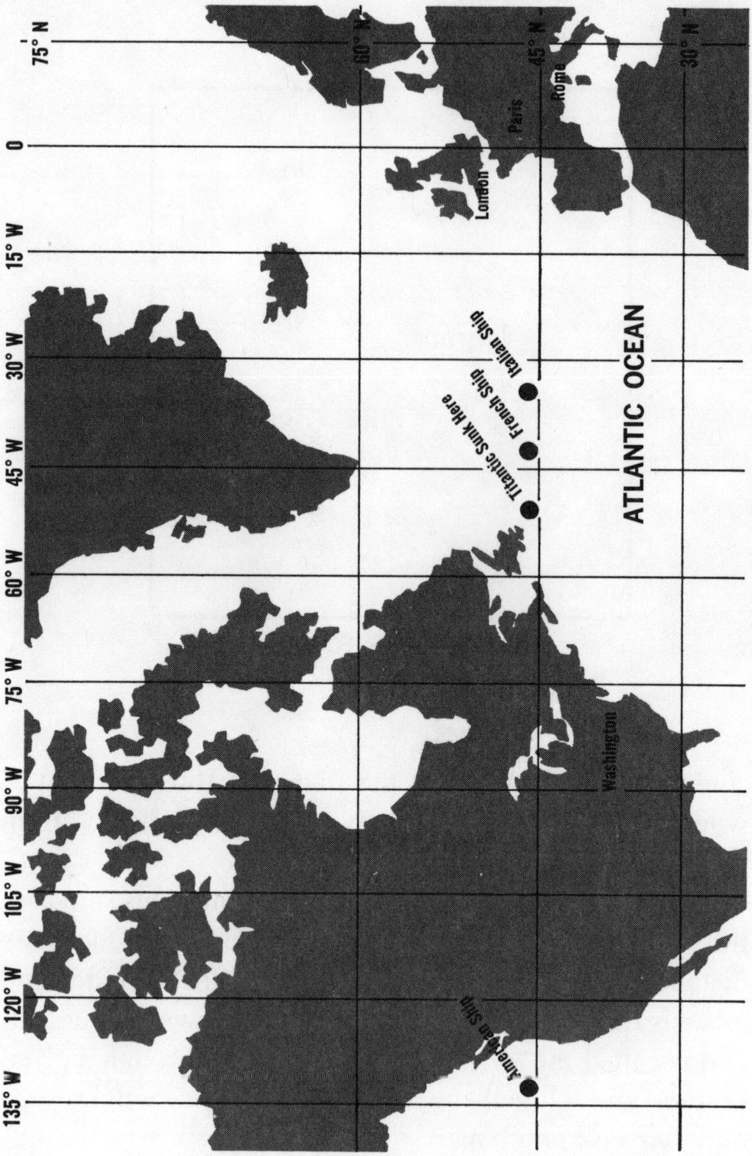
75° N
60° N
45° N
30° N
0
15° W
30° W
45° W
60° W
75° W
90° W
105° W
120° W
135° W
London
Paris
Rome
Italian Ship
French Ship
Titanic Sunk Here
ATLANTIC OCEAN
Washington
American Ship

tance, there must be a starting place. Choosing a starting place to measure latitude was easy. Parallels of latitude become smaller as they approach the poles, and the largest parallel, the equator, is exactly halfway between the poles. There is no such natural starting place for longitude. All the meridians are the same length. And, since there are no east-west poles, there can be no mid-point. But 45 degrees West Longitude has no meaning unless you know west of what, and so map-makers have had to choose a fixed starting place, or *prime meridian*.

Of course, each country wanted the honor of having its capital located on the prime meridian. There are old maps which show the prime meridian passing through Paris, Rome, Washington, and many other cities. On English maps, it passes through the Greenwich Observatory in southwest London. This was confusing enough in olden days, but imagine what would happen if we still used all these cities. When the *Titanic* radioed its position as 50°14′ W. Long. only another English ship would have located it correctly. A French ship would have searched the ocean 60 miles east of the accident because Paris is 60 miles east of London; an Italian ship would have looked for the *Titanic* 500 miles east of its location; while a ship using American charts would have hurried to a spot in the Pacific Ocean off the west coast of Canada.

As England became the most important sea power, more and more maps were made in London. These were so good that they were used by sailors of many nations. Since English maps measured longitude from the Greenwich meridian, it came to be widely used as the prime meridian, and, in 1884, most of the countries of the world agreed to accept it. All places along this meridian from the North to the South Pole are located at 0 degrees Longitude. Places west of the prime meridian are measured in degrees west, for example, New York City is 74° West Longitude. Places east of the Greenwich meridian are measured in degrees east. Berlin is 13° 25′ East Longitude.

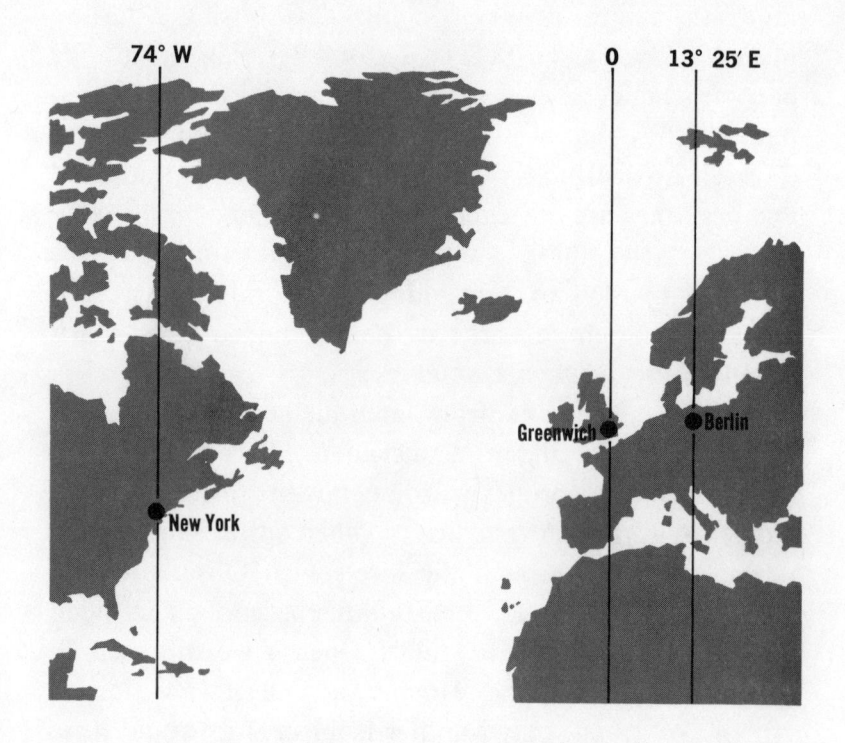

The English map-makers could have measured longitude in only one direction, for instance west. Then Berlin would have been 346° 35′ West Longitude. But instead, they measured halfway around the earth westward and halfway eastward. That is why the 180th meridian is both 180° East Longitude and 180° West Longitude.

It is simple enough to find the longitude of a city on a map because the meridians are drawn and the degrees of longitude are printed across the top and bottom. But a sailor far at sea cannot find his longitude so easily. There are no cities or other visible landmarks on the trackless ocean. Just as he looks to the skies to find his latitude, a sailor can determine his longitude from the sun and the stars.

The men who lived in ancient times learned many things about the earth from studying the heavens. Since the appearance of the sun in the sky divided day from night, it was natural for them to measure time according to the position of the sun. Time measured this way is called *Solar Time*.

If you chart the position of the sun for one day it seems to cross the sky in a great arc, rising out of the eastern horizon at dawn and sinking below the western horizon at dusk. The highest point in this path is called the *zenith*. When the sun reaches its zenith, it is noon Solar Time.

NOW TRY THIS

It is the earth spinning around its axis which makes the sun appear to move across the "bowl" of the sky. Use a flashlight to represent the sun. Hold the light at the level of the equator several feet from a globe of the earth. It is noon all along the meridian which passes through the center of the beam of light. Now turn the globe slowly from west to east. This is the direction in which the earth spins. You can see that it is noon along only one meridian at a time and that as the earth turns, noon occurs farther west.

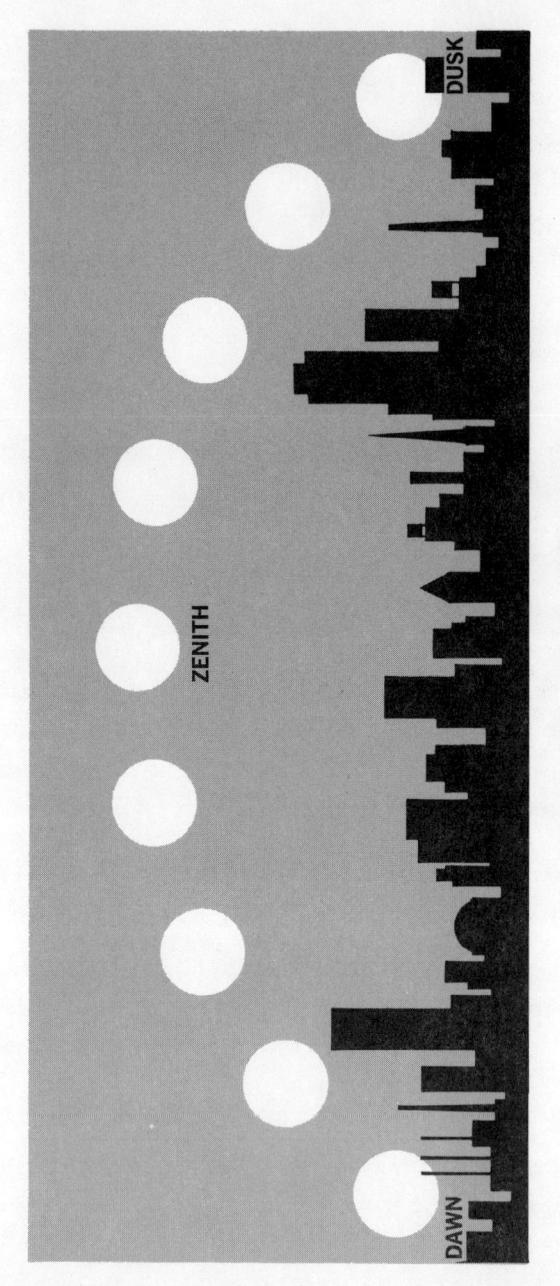

PATH OF THE SUN DURING A DAY

Noon in Chicago comes after it occurs in New York, and noon in San Francisco comes after noon in Chicago. In one hour the earth turns $1/24$ of a complete circle. Since there are 360 degrees in a circle, we know that with the passing of each hour the zenith of the sun will be over a meridian that is $1/24$ of 360 degrees, or 15 degrees, farther west.

The sailor can use the spinning of the earth to measure longitude. By shooting the sun with his sextant, he can find the exact moment when the sun reaches its zenith. This is noon Solar Time all along the meridian on which his ship is located. He can compare this with Greenwich Time, which is Solar Time along the prime meridian and is often written as G.T.

Each hour before noon Greenwich Time is equal to 15° East Longitude, and each hour after noon Greenwich Time means 15° West Longitude. When it is noon on a ship at sea, while back in Greenwich it is 1:00 P.M., the ship's longitude is 15° West. If it is 3:30 P.M.G.T., the ship's longitude is three-and-a-half times 15 degrees, or 52° 30' West. At 11:00 A.M.G.T., the ship's longitude is 15° East, and earlier in the morning, at 8:30 A.M.G.T., the ship would be located somewhere along the meridian of 52° 30' East Longitude. At night, the sailor can shoot the stars and then use the tables in the *Nautical Almanac* to find his longitude.

Even in ancient times, men knew that longitude could be measured by comparing Solar Time with the time at the prime meridian. Unfortunately, this fact was of no use to a sailor until he had an accurate way of always knowing what time it was at the prime meridian. The earliest timepieces, such as the sundial and the hourglass, could not help him. Although some clocks were in use before the time of Columbus, these were tremendous weight-driven machines, each of which needed a whole tower for its case. By the early 1500s watches known as Nuremberg Eggs were being used. Unfortunately, these were often wrong by as much as 16 minutes a day. Since 4 minutes of time is equal to 1 degree of longitude, by the end of a day, a sailor using such a watch might locate his ship as much as 4 degrees from its actual position. At the equator, this would be 276½ miles.

Although there was no accurate method of finding the longitude of a ship, great voyages of discovery were made during the fifteenth, sixteenth, and seventeenth centuries. These explorers used a method of navigation which is called *dead reckoning*. At regular times during the day, a sailor recorded how fast the ship was sailing and in what direction it was moving. By marking on a chart the path of the ship and the distance it had traveled, he could see at a glance the approximate location of his ship. Columbus used dead reckoning when he sailed from Spain in 1492. In later voyages, he was able to sail directly to the same islands by use of this system.

But in the stormy seas of the North Atlantic it is often difficult to keep a ship on an exact course and to determine distance correctly. An error of a few miles might mean little in the middle of the ocean, but this same error could wreck a ship as it approaches a fog-bound coast. Because of her great sea trade, England had to find an accurate way of determining longitude.

MAPPING THE SEA: FINDING THE LONGITUDE

In 1714, the "Commissioners for the Discovery of Longitude at Sea" offered a prize equal to about $100,000 to anyone who could solve the problem. The prize was won, not by a great scientist or scholar, but by a Yorkshire carpenter, John Harrison, who invented the *chronometer* (cro·nom'eh·ter). This is a very accurate watch which is not affected by the pitching and rolling of the ship. It is set to Greenwich Time and no matter where a sailor is, he only needs to look at the chronometer to find the exact time along the prime meridian.

On modern ships, the chronometer and the compass are considered equal in importance as instruments of navigation. The care of the chronometer is the responsibility of one crew member throughout the voyage. Although it can run for 56 hours between windings, the rule is that this crew member must wind it every 24 hours and then report to the captain that the chronometer has been wound. As an additional safety measure, large ships often carry several chronometers.

Most clocks and watches are not set to Solar Time. Since the earth is constantly spinning, the longitudinal line along which it is noon is always changing. For this reason, Solar Time cannot be used conveniently for everyday purposes. Imagine the confusion there would be if watches were set by Solar Time. Time would change whenever you traveled a few miles east or west. When it is 12 o'clock at Beaumont, in eastern Texas, it is 11:11 A.M. Solar Time in El Paso, on the western border of the state; and each city from Beaumont to El Paso has a different Solar Time.

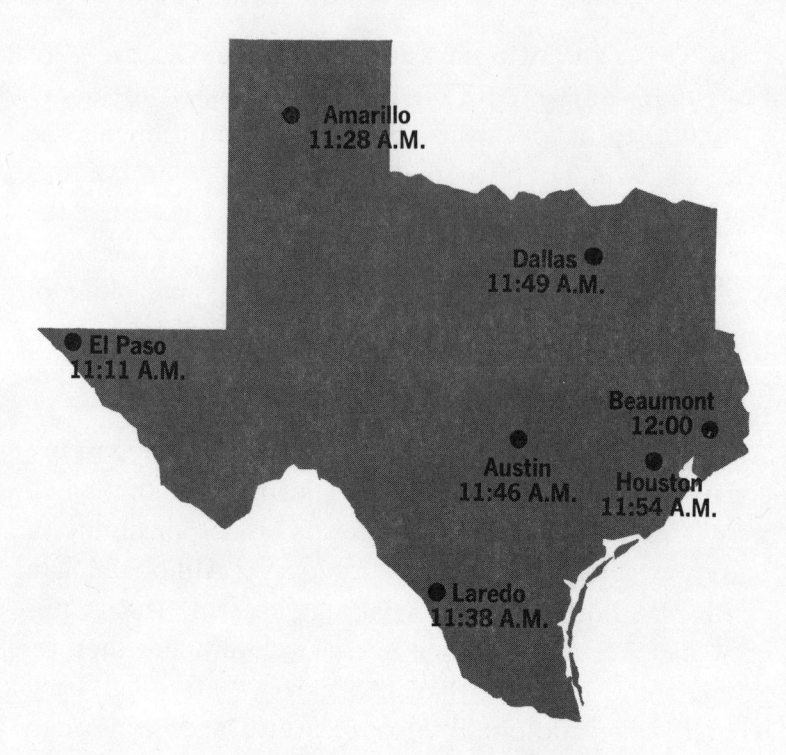

MAP OF TEXAS SHOWING SOLAR TIME

To avoid such confusion, the world has been divided into time zones. A time zone covers roughly 15 degrees of longitude. Standard Civil Time for the entire zone is the same as Solar Time along the meridian which runs through the center of the time zone. Each time zone extends from the North Pole to the South Pole. Rio de Janeiro is in the same time zone as Godthaab, Greenland; Lima, Peru, and New York City are in the same zone; and Cape Town, South Africa, has the same Standard Civil Time as Moscow, U.S.S.R.

When it is 3 A.M. in San Francisco, it is 6 A.M. in New York, and 12 noon at Greenwich. Halfway around the

MAP OF THE WORLD SHOWING TIME ZONES

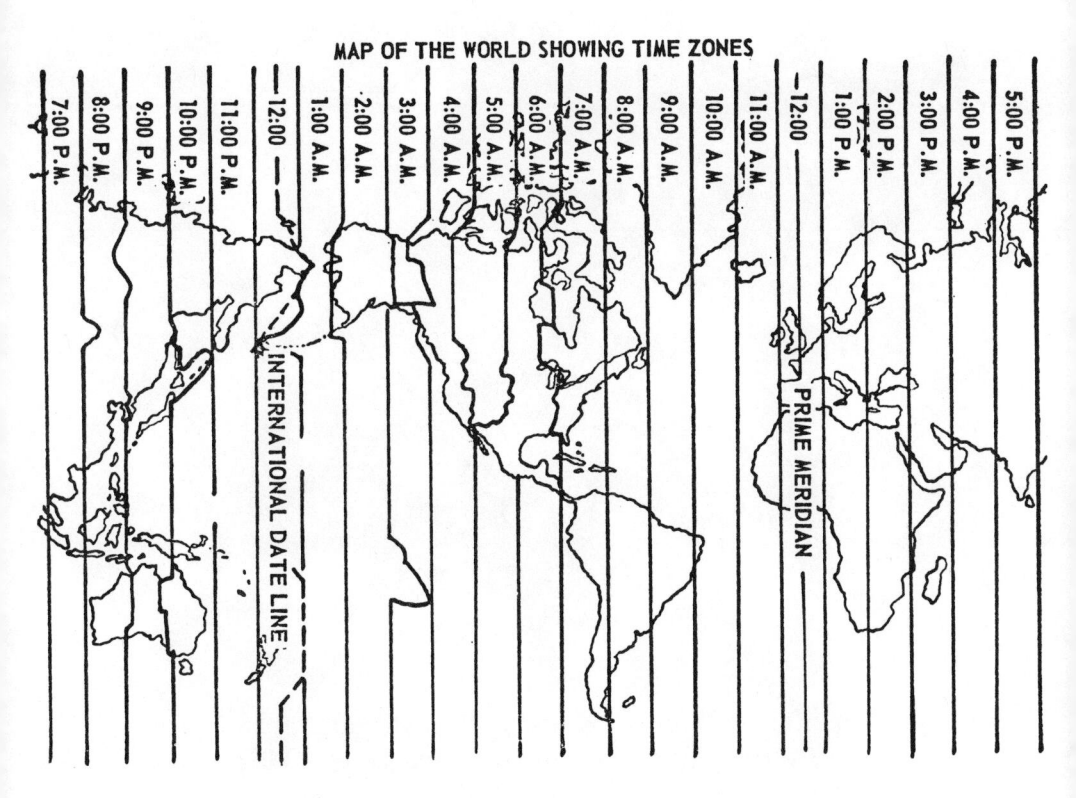

world from Greenwich, along the 180th meridian, it is midnight. The 180th meridian is called the *International Date Line,* for when it is one minute after 12 noon on Monday in Greenwich, on the 180th meridian it is one minute after 12 midnight, and therefore it is already Tuesday.

Dividing the time zones along the meridians works very well at sea. Unfortunately for this system, when men built cities they were not thinking about meridians. As a result,

TIMES ZONES OF THE UNITED STATES

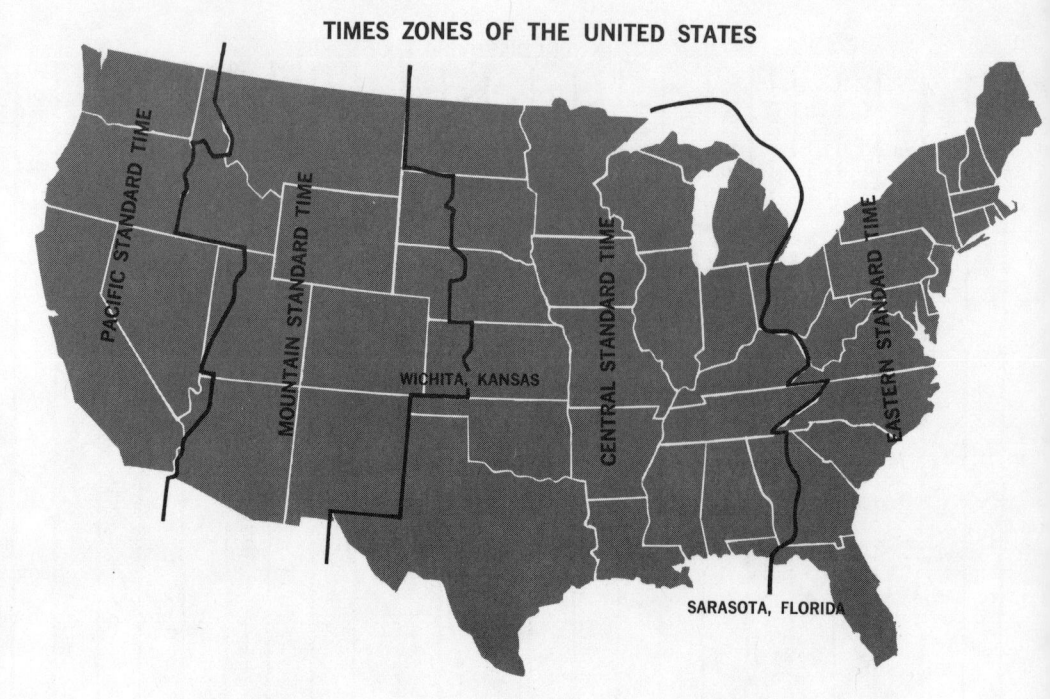

such cities as Wichita, Kansas; Butte, Montana; and Sarasota, Florida, would lie in two time zones. This would make life very difficult. It would be even more difficult on islands which are cut by the meridian of the International Date Line. For convenience, time zones on land do not always follow the meridians exactly.

NOW TRY THIS

You can use Solar Time to find the longitude of your house. Even if you do not have a sextant, there is a way to determine when it is noon Solar Time. You will need two pencils, a sheet of paper, an eraser or cork, a watch set accurately by radio time, and a sunny day. About 11:00 A.M.

MAPPING THE SEA: FINDING THE LONGITUDE

(if it is Daylight Saving Time, start at 12 o'clock) place the paper in the sun on a flat surface such as a sidewalk. Be sure to choose an out-of-the-way place, as the paper must not be moved until the experiment is finished. If it is windy, anchor the paper firmly with something heavy. Push the pencil into the cork so that it will stand straight up. Place the pencil so that it casts a shadow on the paper. With another pencil, mark the end of the shadow on the paper and note the exact time. Mark the shadow end and time every 4 minutes.

In the diagram you can see what happens to the shadow of the pencil as the sun approaches its zenith. When the sun is at its highest point, the pencil shadow is shortest, and it is noon Solar Time. Even the ancient people knew that the shortest shadow occurs at noon.

You probably have no chronometer to tell you Greenwich Time, but if your watch is set accurately and you

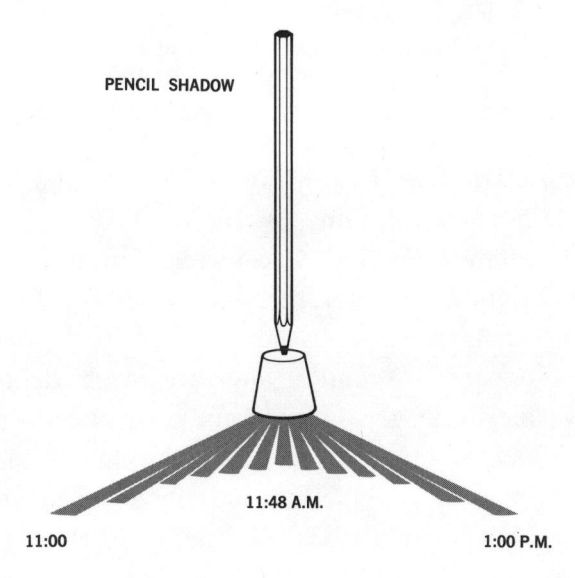

UNDERSTANDING MAPS

know the time zone in which you live, you can find Greenwich Time by using the next table.

TABLE FOR FINDING GREENWICH TIME

Time Zone	Greenwich Time
Atlantic Standard Time	add 4 hours
Atlantic "Summer" Time	add 5 hours
Eastern Standard Time	add 5 hours
Eastern Daylight Time	add 6 hours
Central Standard Time	add 6 hours
Central Daylight Time	add 7 hours
Mountain Standard Time	add 7 hours
Mountain Daylight Time	add 8 hours
Pacific Standard Time	add 8 hours
Pacific Daylight Time	add 9 hours
Alaska	
Juneau	add 8 hours
Fairbanks and Anchorage	add 10 hours
Nome and Aleutian Islands	add 11 hours
Hawaii	add 10 hours
Guam	subtract 10 hours

Suppose you live in Bowling Green, Indiana, and you find that noon Solar Time occurs at 11:48 A.M. Central Standard Time. To find Greenwich Time, you add 6 hours. The time back in Greenwich would be 11:48 plus 6 hours or 5:48 P.M.

Noon occurs 4 minutes later for each degree you travel farther west. If you know the difference in minutes between local noon Solar Time and the time at the prime meridian, you can divide this number by 4 to find your longitude. In Bowling Green, the difference between

MAPPING THE SEA: FINDING THE LONGITUDE

noon and Greenwich Time is 5 hours and 48 minutes. Since there are 60 minutes in an hour, in 5 hours there are 300 minutes and in 5:48 there are 348 minutes. Three hundred forty-eight divided by 4 equals 87. Because the Central Time Zone is west of the prime meridian, the longitude of Bowling Green is 87° West.

The meridians of longitude are numbered on the top and bottom of a map. After you have worked out your longitude, compare your answer with the longitude of your town as shown on a map. It is possible you may find an error of a few degrees. Remember that each 4 minutes, noon occurs one degree westward. If your watch is not set accurately, or if you do not observe the exact minute when the pencil's shadow is shortest, there will be an error. For best results, repeat this experiment on several sunny days.

It is convenient to think of the earth as perfectly round, like a ball or sphere. If this were true, all you would need to know to find the exact distance between any two places is the latitude and longitude of each. However, the earth is not a perfect sphere, and its real shape is still the subject of much study.

For many years, it has been known that the distance around the earth, the *circumference,* is more than 40 miles longer when measured around the equator than it is when measured around a line passing through the poles. Recently, scientists have used satellites launched from the earth to study its shape. They have found the earth to be slightly pear-shaped, that is, fatter toward the South Pole. They have also found that the equator is not a circle, as everyone had supposed, but slightly oval in shape. In determining the shape of the earth, irregularities in the earth's surface, such as mountains and ocean depths, are ignored. However, five low areas and four high areas which affect the earth's shape have been mapped. Perhaps the earth should be described as round, like a slightly battered ball.

Although the use of satellites in mapping the earth is new, the basic method, triangulation, is very old indeed. Like the hovering helicopter, the satellite marks the position of one angle of the triangle. Tracking stations on earth provide the other points. To determine precisely where the satellite is at any given second, the men in the tracking stations use cameras and star charts which show the locations of 260,000 stars. With this information, computers can determine mathematically the position of the ground station. As a result of the United States National Geodetic Satellite Program, a global triangulation network that will have 36 control points is being established. Ground stations have been designed that can be moved by airplane, helicopter, ship, and landing barge to any place where they are needed. Thus, it will be possible to map the entire earth with an accuracy equal to that found on maps of the United States.

Satellites and computers can also be used to replace the sextant and chronometer for determining the locations of ships at sea. American scientists, plotting the course of the Russian *Sputnik* launched in 1957, discovered that its position could be determined from the intensity of the radio signals received from a transmitter on the satellite. The radio signals grow in intensity as the satellite approaches; the signals fade as the satellite travels on.

The first *Sputnik* moved from east to west around the earth, and so was within radio range of only a limited band of earth. The scientists reasoned that a satellite could be placed in a north-south orbit so that it passed over both poles each time it circled the earth. Then as the earth rotated below the satellite's orbit, the satellite would pass within radio range of all areas of the earth.

In 1964, three *Transit* satellites, weighing less than 100 pounds each, were placed in polar orbits about 500 miles above the earth. Four tracking stations from Maine to Hawaii plot the courses of the satellites and determine the exact position each will be in every minute of the following 12 hours. This information is then radioed to the satellite, which in turn broadcasts its exact position every 2 minutes. A radio receiver aboard a ship at sea picks up the signals from a satellite and measures their intensity. A computer aboard the ship uses the intensity measurements and the information broadcast from the satellite to figure the ship's position within an error of only a few yards. The ship does not have to wait for the next orbit of the satellite to determine its position again. Because there are three such satellites, the computer can use the information from one of the other two as it passes overhead.

In fact, with modern equipment, it is even possible to replace the sailor; a ship could sail from one port to another without a man aboard. But the necessary instruments are still so expensive and bulky that they are unlikely to replace sailors very soon. And for people who sail the seas for pleasure and adventure, such instruments are unlikely ever to replace charts.

Mapping the Sea: Nautical Charts
7

Long before men came to live on the islands that cluster at the mouth of the Hudson River, a great glacier moved down from the north. Huge boulders were carried by the glacier as easily as specks of dust are carried by the wind. As the glacier moved forward, some of the boulders crashed together and were ground into small stones. Other boulders moved majestically along, smoothed but hardly reduced in size. In time, the climate changed and the edges of the glacier melted. The snow and ice disappeared, but the boulders remained, some of them scattered throughout what is today Long Island Sound.

Now the glaciers have gone. People have come to the islands at the mouth of the Hudson, and they use boats and seaplanes for traveling. The navigators of these boats and planes need many charts to help them reach their destinations.

Sailing charts showing the latitude and longitude of the land masses are the road maps of the ocean. They are used to sail between distant ports. But as a ship approaches land, *harbor charts,* which are carefully marked to show the depth of the water in the harbors, and *shore charts,*

MAPPING THE SEA: NAUTICAL CHARTS

which show the depth along the coast, are important. Special charts which show both ocean depths and obstacles are prepared for the owners of small boats.

Although the bottom of the ocean is called its floor, it is unlike any floor you have ever seen. Near the coast there may be boulders and gravel, the remains of a glacier. Farther out at sea, rising from the ocean bottom, are mountain peaks taller than many mountains on dry land, and yet they do not break the surface of the water. The tallest mountain in the world is not Mt. Everest, which at the latest survey measured 29,142 feet above sea level; it is Mauna Kea in Hawaii. Mauna Kea is the peak of a volcanic island and rises 31,000 feet from the floor of the Pacific Ocean. But only the top 13,796 feet can be seen above the water. The sea has canyons which are deeper and more impressive than the Grand Canyon of the Colorado River. Cliffs almost 5 miles high rise in the middle of the Atlantic, and yet they are still hidden by a mile of water.

Under a ship the ocean floor may have a rug of gently rippling sand. It may be strewn with huge boulders or be covered with many feet of thick, slimy ooze. The depth of the ocean ranges from its shores, which are covered only at high tide, to the Challenger Deep near the Island of Guam in the Pacific, where it goes down almost 7 miles.

Charts of the ocean bottom are part of the equipment of every ship. Even though it is made of the strongest steel, a ship will not last long if it is pounded against a reef. A ship stuck on a sand bar is of no use to its owners. Maps warn the sailor of such dangers ahead.

The peaks, deeps, and reefs also serve as landmarks.

UNDERSTANDING MAPS

Just as you use mountain peaks and rivers on land, the sailor can check his location by comparing the pattern of the ocean floor with his charts. On land only your eyes are needed to find landmarks, but the sailor needs more than his eyes to find the landmarks of the ocean floor. He must take *soundings*. The oldest and simplest device is the *hand lead*. This is a lead cone with an eye in the top to which the line is attached. In the bottom is a hole which can be filled with tallow or grease whenever a sample of the ocean bed is needed. When the lead touches bottom, bits of the sea floor stick to the grease.

The line on the hand lead is usually made of braided flax. It is marked in *fathoms*. One fathom equals 6 feet. Since the line is often used at night, it has markers so that the sailor can "read" the line by feel. Standard markers are shown on the chart below.

Fathoms	*Markers*
2	2 strips of leather
3	3 strips of leather
5	white cotton rag
10	piece of leather with a hole in it
20	a cord with 2 knots
25	a cord with 1 knot
30	a cord with 3 knots
35	a cord with 1 knot
40	a cord with 4 knots
45	a cord with 1 knot

As a sailor "reads" the line, he calls out, "By the mark three," "By the mark five," or whatever depth he finds. The famous American writer, Samuel Clemens, who once was a pilot on the Mississippi River boats, chose his pen name from the 2-fathom report of the sailors who manned the hand leads, "By the mark twain."

For deep-sea soundings, the *dipsey lead* can be used. This may weigh from 30 to 100 pounds, and its line is made of piano wire. In very deep water, it takes several hours to make one sounding by this method. For this reason, the dipsey lead is seldom used today. Modern ships use a *fathometer,* one of the first electronic surveying devices developed. The fathometer sends a ping of sound down through the water. When the sound strikes the ocean floor, the echo bounces back to the ship. The fathometer keeps an accurate record of the length of time it takes the sound to travel down and back. Sound travels faster in water than it does in air. It travels about 800 fathoms each second. If it takes one second for the ping to reach bottom and return, the depth of the ocean floor is one-half of 800 fathoms, or 400 fathoms. With a fathometer it takes only 6.6 seconds to measure a depth of 3 miles: 3.3 seconds for the ping to travel 3 miles down and 3.3 seconds for it to travel 3 miles back to the ship. Imagine lowering a 100-pound dipsey lead on 3 miles of piano wire!

The fathometer automatically records the depths on a *fathogram,* by drawing a graph which shows the contours of the ocean floor. There is a picture of a fathogram on page 98. Deep-sea fathometers can measure a depth of 4½ miles with an error of less than 60 feet. Smaller fathometers are used in the shallower waters found a few miles from the coast line. Here the error is not greater than 2 feet. The Coast and Geodetic Survey uses fathometers to sound about 50,000 to 75,000 square miles of ocean floor every year.

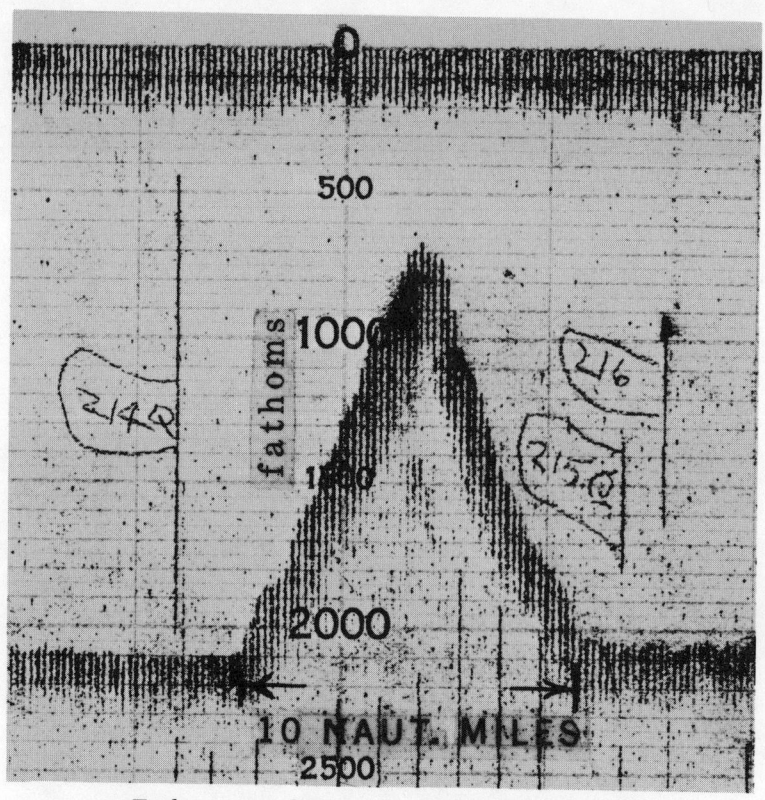

Fathogram of an underwater volcanic peak

Close to shore, a simple but accurate device can be used. A wire is suspended between two ships at a certain depth, for example, 3 fathoms. The ships sail back and forth dragging the wire between them 3 fathoms below the surface. Any obstacle which sticks up above 3 fathoms snags the wire, and the position of the obstacle is then charted. After the entire area has been charted at the 3-fathom depth, the level of the wire is changed, and the charting is continued for the new depth. Wire-dragging was used formerly to chart the positions of the boulders on the floor of Long Island Sound. At present, high-speed electronic sounding devices and shipboard computers are used.

98

The Coast and Geodetic Survey usually wire-drags about 200 square miles of coastal waters each year. Before a wire-drag survey begins, every effort is made to inform the local citizens of what is planned. In spite of this, the Survey ships have sometimes been shot at by angry fishermen who think they are chasing the fish and by lobstermen who fear they will upset their lobster pots. Treasure hunters, on the other hand, welcome the wire-drag teams because their work may reveal the location of an ancient wreck. The Coast and Geodetic Survey teams do not try to identify the wrecks they find; they simply locate them on the shore charts. They leave identification to treasure-hunting scuba divers.

Sailing charts of the North Atlantic, South Atlantic, North Pacific, South Pacific, Indian Ocean, and Central American waters are published regularly by the U.S. Hydrographic Office. New charts are published frequently, but in the meantime, the older editions are hand-corrected. It has been said that the only possible use for an out-of-date sea chart is to explain why the wreck occurred. The new charts are necessary because, in addition to information about depths and soundings, these indicate the winds and currents that can be expected.

Currents move through the ocean in somewhat the same way as rivers and streams move across the land. The drops of water which wash the ocean beach this year are not the same drops which splashed there a year ago. Some of the water in which you swim may once have sparkled on the beaches of tropical islands. Other drops may have helped to float icebergs as they slipped from the glaciers of Greenland. The waters of the ocean are never still. They are

MAIN ATLANTIC OCEAN CURRENTS

being mixed and churned constantly by the swiftly moving currents. Because the course of the currents is more or less fixed, they can be shown on charts.

The currents of the ocean are created by many different forces. The winds constantly blowing over the surface, the rotation of the earth, the changing temperatures, and the differences in saltiness of the waters all play a part. Certainly you have noticed the difference in temperature of the air in a room. On the floor you may feel cool, but if you climb a step ladder so that your head is near the ceiling, you will feel much warmer. Cold air is heavier than warm air, and cold water is heavier than warm water. Icy water from the Arctic and Antarctic regions slips downward into the ocean depths, while water from the seas near the equator forms the warm currents which flow at the surface.

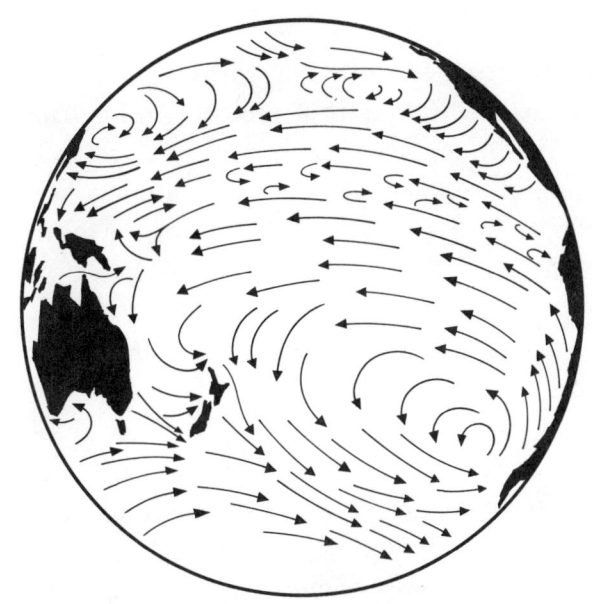

MAIN PACIFIC OCEAN CURRENTS

In the same way, the heavy, saltier water from the Mediterranean Sea sinks to the bottom and flows into the Atlantic deeps, while a current of lighter, less salty water floods back to take its place.

There is a two-way traffic of more salty and less salty water passing through the Strait of Gibraltar. During World War II, German submarines used these currents to help them sneak past the guns of the British fort at Gibraltar. With their engines dead, they slithered in with the less salty upper current, and out again in the saltier lower current.

Regular ships also need information about the currents. Unless the sailor is prepared, strong and tricky currents in harbors and along the shores can quickly wreck his ship. Powerful ocean-going currents such as the Gulf Stream help the sailor by speeding his ship across the ocean.

The first person to study the warm Gulf Stream waters scientifically was Benjamin Franklin. On his trips across the Atlantic to represent his country in Europe, Franklin took temperature readings of the Gulf Stream and the surrounding ocean. He also recorded his observations of the color of the water and the tropical plant life which he found in the Stream. He recommended that ships going to Europe should ride the Stream, while those coming to America should cut sharply across it. But the ships' masters were an independent group and refused to follow the notions of a landlubber. There is, however, reason to believe that some of the ships which made the voyages in record time accidentally did what Franklin recommended. The navigators of modern steamships know more about the Gulf Stream, and they do take advantage of it.

The first official study of the Gulf Stream was made under the direction of Franklin's great-grandson, Professor Alexander D. Bache, in the summer of 1846. Professor Bache was the second superintendent of the United States Coast Survey, which later became the Coast and Geodetic Survey. The study of the Gulf Stream still continues. Where Franklin once lowered his own thermometer, the Coast and Geodetic Survey now sends its floating laboratory ships, which still study the temperature of the Stream.

The behavior of the Gulf Stream is vital to the climate of the East Coast of the United States and to Great Britain and Ireland. Dublin and London are farther north than Newfoundland, yet their winters are milder than those of New York City because of the path of the Gulf Stream. A change in that path bringing the warm waters of the

Stream closer to New York would result in much warmer winters for that city.

Not all ocean currents are warmer than the surrounding waters. The California current, which flows off the West Coast, is colder than the adjacent waters. Moisture-laden air turns to fog as it passes over this current, and almost daily the fog moves in across the Oregon and northern California coasts. Thus the California current provides the moist climate without which the giant redwood trees could not survive.

Information about the ocean currents is very important to fishermen. The kind and quantity of fish found in any location depend in part on the temperature of the water. On the other hand, scientists use information about the animal and plant life to determine the boundaries between the oceans. The Atlantic and Arctic Oceans are not separated by land in many places, but the Canadian Board of Fisheries has found a way to determine where one begins and the other ends. They have discovered a variety of microscopic animal which can exist only in Arctic waters. Another variety of the same animal can exist only in Atlantic waters. By making microscopic examinations of the water to determine which variety is present, the scientists are able to chart the boundary between these oceans.

The boundaries between the land and the oceans are easier to observe, even though they are constantly changing. On the shore, the water line climbs steadily until it reaches its farthest position; this is high tide. Then the water recedes until it reaches its lowest line, low tide. Most ocean shores experience two high and two low tides a day.

Some charts show high and low tide lines. These charts

are often prepared from photographs. Aerial photographs for charting may be made with black and white film, color film, or infrared film. Black and white film is the least expensive, but color film is more effective in revealing changes in the depth of shallow water. The greater the depth of the water, the darker the color will appear. The color film clearly shows color shading that is not visible even to an observer in a plane. Infrared film shows all water as black. With infrared film, it is possible to see the exact water line at the time the film was exposed. On page 105, there is an infrared photograph of a beach. By timing the flights to the tides, it is possible to determine the high and low tide lines for any day. But that does not mean that the lines will be the same the next day, since the high and low tide lines vary daily. Many nautical charts, like the one on page 106, show the mean (average) low-water lines.

Since the sea is ever-changing, the men who guide the great ships across the ocean need as many aids as possible. During World War II, a new aid to navigation was worked out, and with it came a new kind of chart. The aid is called *loran,* short for LOng RAnge Navigation. There is a series of curves printed over a chart of the ocean. Loran stations, which are radio sending stations, have been built along most of the important coast lines of the world. A loran station is made up of one master and two slave units, each of which sends out radio signals.

A receiving set on board ship measures the difference in the length of time it takes for the radio signals to travel from each of the units of a single loran station to the ship. There is one curve, and only one curve, on the loran chart for each time difference. When this curve has been found

Infrared photograph of Myrtle Beach, South Carolina

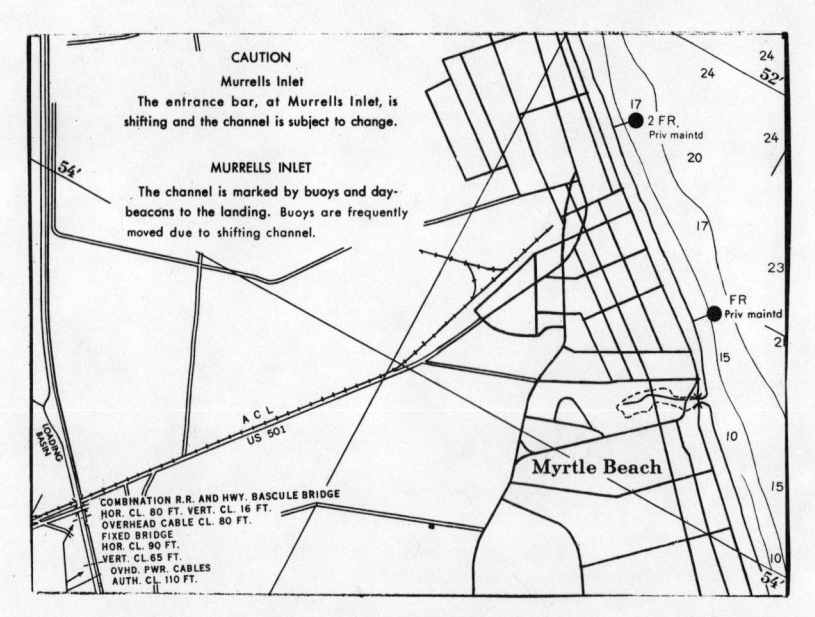

Chart showing mean (average) low-water lines

and marked, the receiving set is tuned to another loran station and again the time difference is measured. The ship's position is the point where the curve of the second station crosses the curve of the first station. With loran, a sailor can find his position in a few minutes almost as accurately as he can by shooting the sun, the stars, or a satellite.

Mapping the Sky: Man Among the Stars
8

Wherever men travel they make paths for others to follow. Everyone on earth has used some kind of road during his life. Whether they are made of dirt, macadam, or concrete, the land roads form a network across the face of the earth. Roads do not wind aimlessly, even though sometimes it seems so. They are built to connect the places which are important to men, and they twist and turn in order to avoid such obstacles as hills, cliffs, and rivers.

Perhaps it is difficult to imagine roads on the sea. Although there are no strips of concrete stretching across the ocean waters, there are roads. The roads of the sea are the shipping lanes. These lanes connect the coastal cities of the world. Shipping lanes are more direct than land roads because there are no hills, cliffs, or rivers in the way. The path of a sea road depends on the ocean currents, the winds, the weather, and such obstacles as reefs and icebergs. Ships rarely travel in the areas of the ocean that are not part of the shipping lanes.

The lanes are wide and often very busy. Transatlantic ships usually sight a number of other ships on each voyage. Formerly, the lanes were not clearly defined, and the ship captains were free to sail in whichever parts of the lanes they chose. In July, 1956, the liners *Andrea Doria* and *Stockholm* were headed in opposite directions off the coast of Nantucket. They crashed in an otherwise empty part of the sea, and 51 lives were lost. As a result, 1,108 definite sea lanes were established.

Today there are also *safeways* in the Gulf of Mexico. The safeways are necessary because of the many floating oil-pumping stations that dot the Gulf. Not only are ships expected to stay within the safeways, but no one is permitted to drill an oil well within the safeways.

There are also roads in the sky. These are not built with millions of tons of concrete. Instead, they are paved with invisible radio beams. Perhaps you live under a sky road. If each day at about the same time, planes traveling in the same general direction pass over your house, you can be sure there is an *airway* over your head. Airways connect the cities of the world. They are usually very direct, but they do depend in part on the winds, the weather, and the presence of emergency landing fields.

In the early days of the airplane, there were no roads in the air. Even after the airplane proved its military value during World War I, there were only a few planes in use. Most of these were used to amuse people at carnivals and country fairs. A pilot could take off from any large, smooth field, fly in any direction he pleased, and land wherever he wished. The air was a trackless frontier like the wilderness that faced the early American pioneers.

At first, the pilots used a system of navigation called *contact flying*. The flyer found his way across the countryside by following landmarks. Contact flying is like the church steeple navigation of the Middle Ages. The pilot must have clear daylight and must remain within sight of familiar landmarks. Today, a contact flight is known as V.F.R., visual flight rules.

The earliest pilots used automobile road maps and U.S. Coast and Geodetic Survey maps for contact flights. Villages, rivers, and especially railroad tracks were used as guides. This worked very well during the day, but at night all planes had to be grounded. As long as planes were used mainly for amusement, night flying was unimportant, but in 1920 the U.S. Government decided to start transcontinental air-mail service.

In the beginning air mail was flown only from dawn to dusk. At the end of the day, the plane landed and the mail was loaded onto a train. All through the night the mail sacks traveled on land; then, the next morning, they were transferred to a plane and continued their trip by air. A letter using this service crossed the country 22 hours faster than formerly.

After one year of this plane-train service, the U.S. Post Office decided it wanted an all-plane service. Some kind of beacon was needed to guide the plane at night. For the first flight, farmers helped by lighting bonfires along the way. Because this flight was a success, Congress voted to spend the money to build flashing beacons to mark the airways of the United States.

Since a beacon has little value unless a pilot knows where it is located, the U.S. Coast and Geodetic Survey, which prepares all non-military flight maps in the United States, produces visual charts for V.F.R. flights. These are topographical maps which show the contours of the land, landmarks such as airports and beacons, and obstacles. Outdoor theaters are usually shown because they are easily seen from the air. Electric power lines are often included because they are both obstacles and obvious landmarks. In wooded areas, the trees are cut away to make room for the lines. Where few other landmarks exist, these long open streaks are important to pilots. Whenever power lines are strung from one hill to another high above a valley, they must be shown so that low-flying planes do not crash into them.

Instrument Approach and Procedure Charts are also prepared for all major airports. These are used for radio contact and instrument approaches in bad weather or whenever the visibility is poor.

Because obstacles such as buildings and beacons are frequently changed or added, the useful life of an aeronautical map is very short. The Coast and Geodetic Survey receives an average of 80 notices of changes a day. These are marked on master charts, and every 28 days corrected maps are printed and mailed to pilots.

At regular intervals along the modern American airways, radio transmitting stations send out constant signals in certain directions. One antenna sends the International Code for *"A"* (dot-dash) while another sends the code for *"N"* (dash-dot). Together these signals form a broad beam. The receiving set of a plane flying down one side of a beam will receive only dash-dot. A plane on the opposite side will receive only dot-dash, but a plane in the center will receive both signals. A mixture of dot-dash and dash-dot sounds like a constant hum. When a pilot hears this hum in his ear phones, he knows he is "on the beam," and

LAKE ST. CLAIR

242

WINDSOR RADIO

RADIO RANGE COURSE

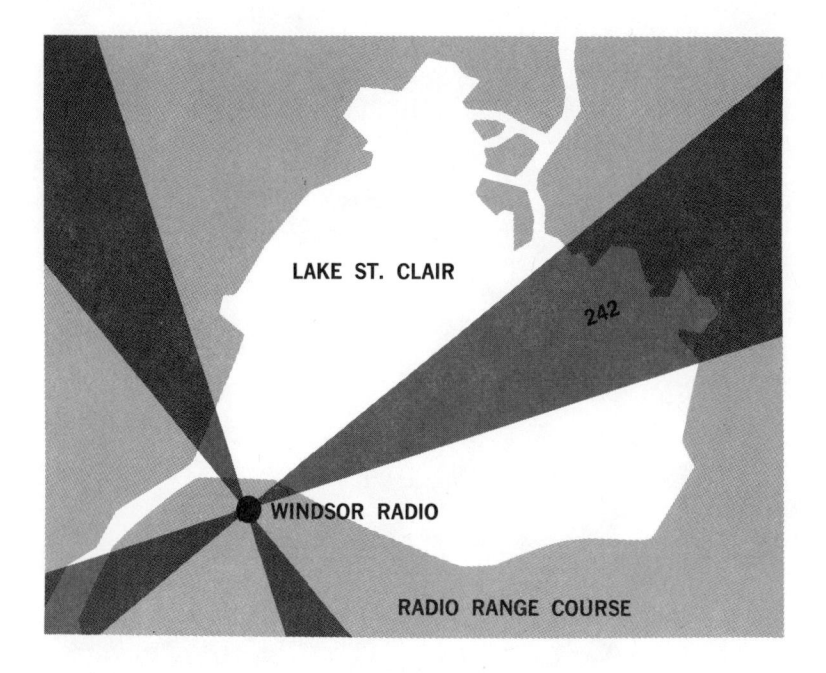

headed straight for his destination.

The beams of *radio range courses* are marked in magenta on *Radio Facilities Charts*. Radio beams fan out as they pass through the air. The farther the beam is from the station, the wider the area it covers. For this reason, the beam is shown on the chart as a giant pointer. The compass directions are marked in degrees on each beam, and each chart has several compass roses for convenience.

The circle of a compass rose, or the card of the compass itself, may be marked in two ways. It may be divided into 360 degrees as on the Radio Facilities Charts. This makes it possible to measure direction very accurately. Or, like early compasses, it may be divided into thirty-two points, each named for a direction. For example, the points between North and East are:

>North
>North by East
>North North East
>North East by North
>North East
>North East by East
>East North East
>East by North
>East

Compass Card

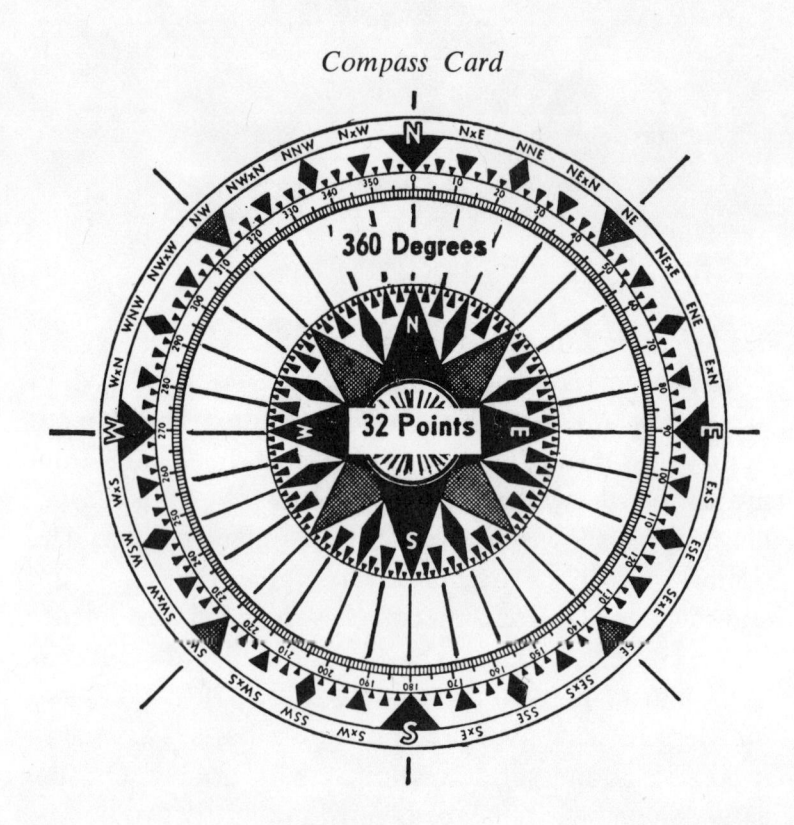

When you have learned the points all the way around, you can "box the compass." Today many compass cards are marked in both degrees and points.

The greatest advantage of airplane travel is speed. Even before the invention of the airplane, men kept increasing the speed with which they traveled on land. A crack train, for example, can speed along at more than 90 miles per hour. Ocean travel, however, has always been slower, and ocean distances are greater. Even today, the fastest luxury liner averages less than 30 miles per hour. But 30 miles an hour is very slow when one is traveling the great distances which separate major ports. It is 2,091 *nautical miles* from San Francisco, California, to Honolulu, Hawaii, and 6,221 nautical miles from San Francisco to Manila in the Philippine Islands. (A nautical mile is 6,076.1 feet compared to 5,280 feet in a statute or land mile.) The distance from New York City to Cherbourg, France, is 3,154 nautical miles, and it still takes more than four days to make the crossing by ship.

With the first successful flight, people began to dream of using aircraft to span the ocean. But it was many years before this dream could come true. It is much more difficult to plot a plane course over the water than over the land. Contact flying is impossible, for there are no visible landmarks. Moreover, a plane must often fly through heavy fog or at night. It certainly would not be practical to try to build a string of beacons across the ocean.

There are so many problems involved in airplane navigation over water that many people believed it would never be possible, but fliers continued to dream of crossing the ocean. When Charles Lindbergh planned to cross the

Atlantic in 1927, he purchased regular ship charts from a ship-chandler's shop at San Pedro, California. In addition to maps showing latitude, longitude, and land masses, he bought a time-zone chart and weather maps. Once, as he was flying over Nova Scotia on his way to Paris, the wind caught a corner of a map and fluttered it toward a window. He grabbed it back quickly, for if it had blown away, he would have been forced to return to New York. He was flying on course; his plane was in excellent condition; he had enough gas; but, without his chart, he never could have reached Paris.

Like Columbus, Lindbergh steered his way across the ocean by dead reckoning. He followed the course marked on a ship's chart, using only compasses to indicate the direction and the plane's speedometer to help him estimate the distance. Since he flew alone, he had to check the course as well as handle the controls of the plane. The airplanes that fly the oceans today have a navigator who is responsible for plotting the route and seeing that the plane remains on course.

The airplane navigator, like the ship navigator, depends on dead reckoning, satellites, radio, and the stars. It may seem strange, but the men who guide our most modern means of transportation depend in part on the oldest science in the world, *astronomy*, the study of the stars. Star charts were among the first maps made.

Look up into the night sky and pick out a star. It seems to be only a twinkle of light in a diamond-studded sky. Because the earth is spinning, the stars, like the sun, never seem to stay in the same place. Each hour and each day your special star will appear in a different spot. If you

● DUBHE

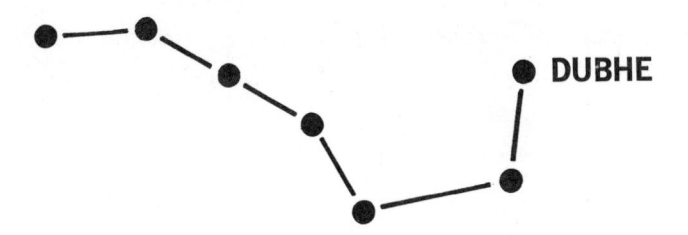

want to study this star, you must have a sure way of locating it each night.

One way to find a star in the heavens or on a chart was worked out by the people of the ancient world. Just as you sometimes can see animals or faces in the clouds, they pictured figures outlined by the stars. To each of these groups of stars or *constellations,* they gave the name of an imaginary hero or an animal, and they made up wonderful legends about them.

Different tribes saw different pictures in the stars. The people of ancient Greece made charts that showed 48 constellations. These are still used on the star charts of the Northern Hemisphere. Suppose the star which you chose is the first one in the pointer of the **Big Dipper** sometimes called Ursa (Er'suh) Major. You can locate this star, which the Arabian astronomers named *Dubhe* (Dew'bee), by first finding the Big Dipper. This is like finding a country on a map when you know on which continent it is located.

NOW TRY THIS

You can plot the course of Dubhe across the heavens. On paper, draw a large circle to represent the "bowl" of

UNDERSTANDING MAPS

the sky. Mark the Pole Star, sometimes called Polaris, in the center of the circle and draw a line from the horizon (the edge of the paper) to the Pole Star. To orient your map, be sure that this line is pointing north each time you chart the position of the star. Now pick out Dubhe in the night sky and mark its position in the circle. Once each week at the same hour, mark its position on your chart. In three months, Dubhe will have traveled one-fourth of the way around Polaris. In six months it will be halfway around. And if you continue for a whole year, you will find that Dubhe is right back where it started.

Today there are other ways to locate stars. If the latitude and longitude of a ship at sea is known, its position can be pinpointed without difficulty. A similar system has

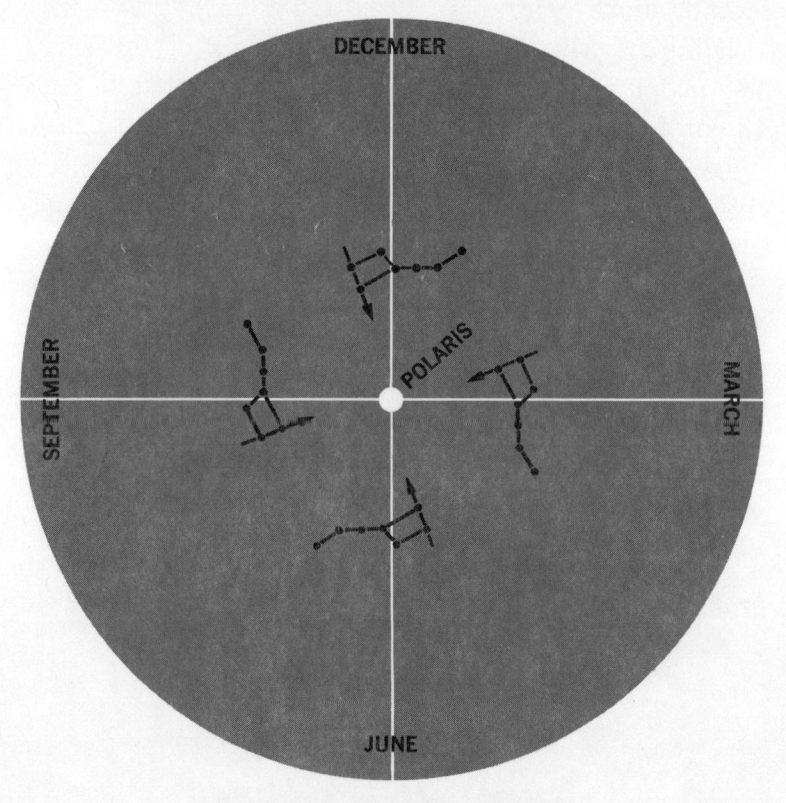

PATH OF DUBHE AROUND POLARIS

been worked out for the sky. Just as there are imaginary poles and an equator on earth, there are imaginary poles and an equator on the "bowl" of the sky. The North Star, Polaris, is just 1 degree from the imaginary north pole. Latitude on the "bowl" of the sky is called *declination* (deck′lin·ay″shun) and longitude is *right ascension* (as·en-′shun). Like latitude and longitude on earth, these may be measured in degrees. The declination of Dubhe is 63 degrees and its right ascension is 175 degrees. If you know the declination and right ascension of a star you can find it on a chart or in the sky even if it is not part of a constellation. This is important today, since we have discovered many stars which can be seen only with a powerful telescope.

A very interesting type of star chart has a movable center card or wheel like the one shown in the next illustration. By turning this, it is possible to see exactly which stars are overhead at each hour of each day and night of the year. Some of these charts show declination and right ascension as well as the constellations.

The positions of the planets are not shown on the movable wheel charts. The astronomers of the ancient world noticed that some heavenly bodies did not stay in a fixed place as the stars wheeled around. They named them *planets* from the Greek word meaning wanderer, since they wandered in and out among the stars during the course of the year. Today we know that they are not stars at all, but are parts of the Solar System like our earth. The planets follow a definite path as they revolve around the sun. The *ecliptic,* the great circle path which they seem to follow through the sky, is often marked on charts. Tables

WHEEL CHART OF THE SKY

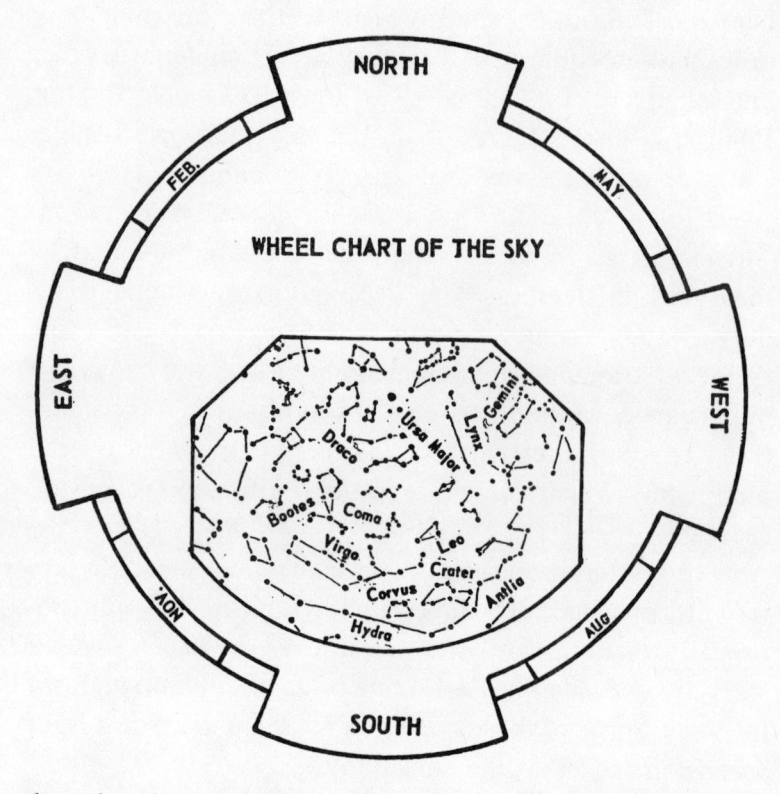

that give the declination and right ascension of the planets for each day of each year make it possible for you to locate them on the ecliptic.

Astronomers are now working on maps of the surfaces of some of the planets. Maps of Mars show the positions of the polar caps and the dark areas where it is thought vegetation may be growing. By comparing old maps and photographs with those made recently, the astronomers learned that a new dark area about the size of Texas has appeared in what was believed to be a desert. An interesting early map of Mars was prepared by Professor Percival

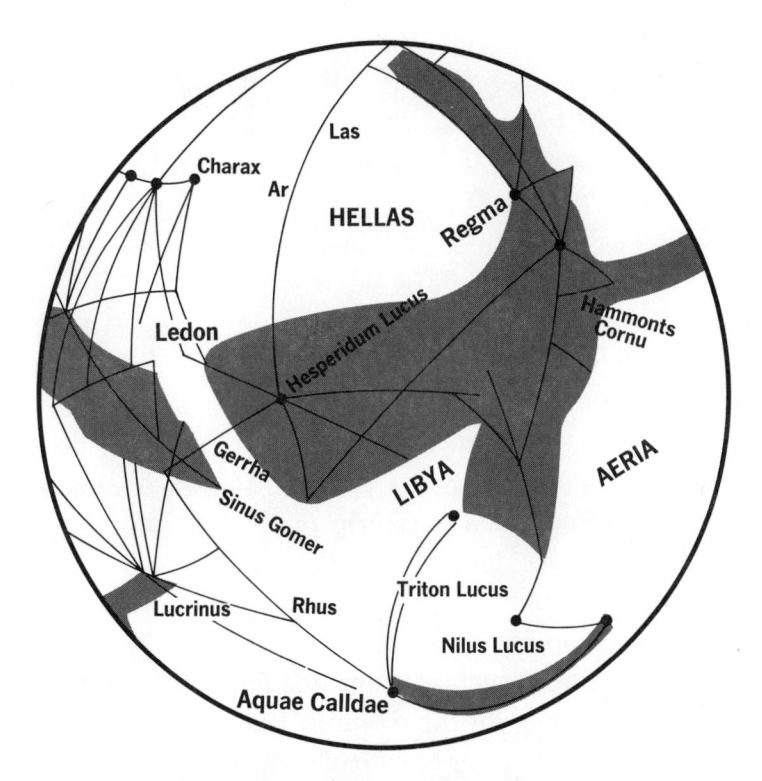

PROFESSOR LOWELL'S MAP OF MARS

Lowell, founder of the Lowell Observatory in Flagstaff, Arizona. A copy of this map appears in the illustration above. It shows the lines or canals which some astronomers have seen on Mars. Professor Lowell suggested that these are irrigation ditches made by intelligent beings. This theory is not accepted generally by present-day scientists.

An amazing project is the mapping of the surface of Venus. Although this planet is earth's closest planetary neighbor, it still takes a rocket about three months to reach it. Moreover, the planet is perpetually covered by a heavy layer of clouds which cannot be penetrated by an ordinary telescope. In spite of these handicaps, it is now known that there are at least two great mountain ranges on the surface of Venus. One range, called Alpha, runs about 2,400 miles north and south and is several hundred miles wide. (The Rocky Mountains extend from Alaska into New Mexico, a distance of about 3,000 miles.) The other range, Beta, runs east and west and is even more extensive. These mountains were discovered by aiming radar beams at Venus and using computers to analyze the returning radar echoes. If the beams strike a flat area, the echoes all return at about the same instant. If the beams are reflected from a mountain range, the echoes from the radar rays that hit the mountain tops return before those which strike farther down the sides of the mountains.

It took very complex instruments to gather information for the first map of Venus, but the first map of the moon was made without such instruments. Galileo Galilei produced the first map of the moon in 1610, making his observations through a very crude telescope that was nothing more than two simple magnifying glasses. By studying the shadows cast by the lunar mountains and using triangulation, Galileo estimated the heights of the mountains. He also noted the dark areas, which for many years were assumed to be oceans. They were named *mares,* from the Latin word which means seas. In the years since Galileo's original work, there have been many maps of the moon. Man has learned that the lunar seas contain no water, and many new details have been added.

Photographs of the moon, taken through huge modern telescopes, have been used to make photogrammetric measurements and to produce topographical maps of the

Photograph of the moon's Tycho crater taken by Orbiter V

moon like the one on page 122. More recently, cameras mounted on rockets have sent back close-up photographs of the moon's surface. Moon maps are essential for moon exploration. For example, the keyhole-shaped crater Fauth is only 13 miles wide, but it is about 4,500 feet deep. This is certainly no place to land a space ship. Other areas are strewn with volcanic rocks which make landing impossible. Because of the harsh conditions on the moon, no air, and great daily changes in temperature, moon explorers cannot wander around looking for a mountain pass as men did on earth when they were exploring America. Moon explorers must know where they are going and

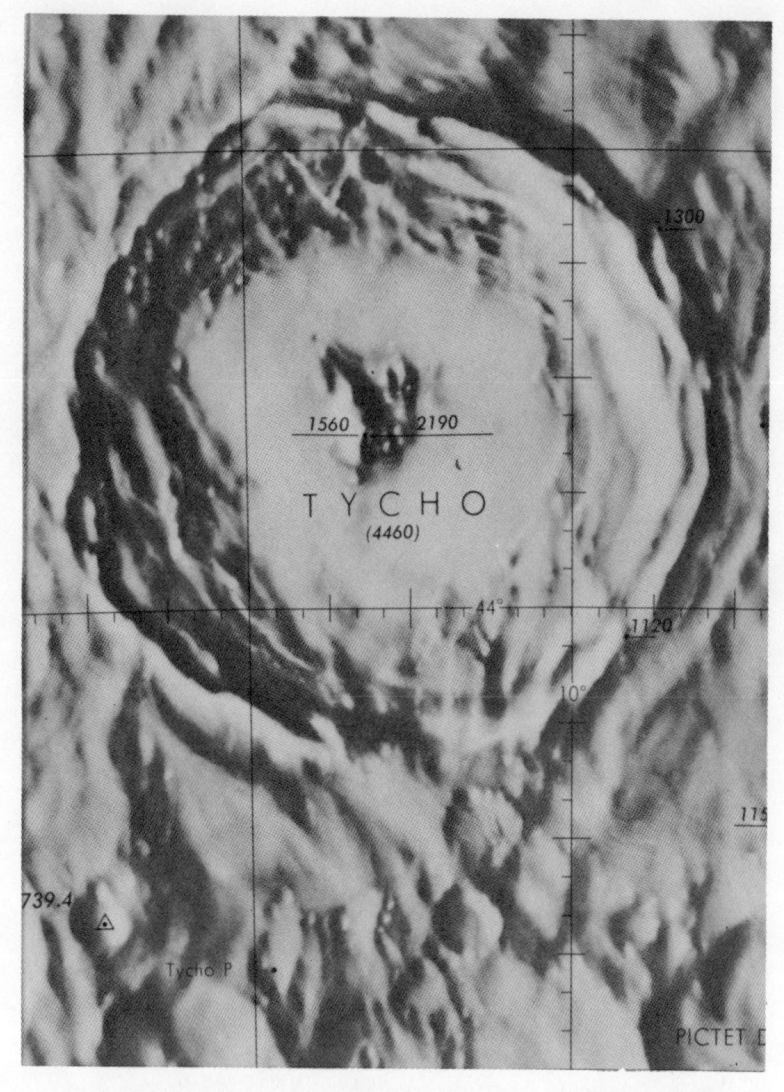

Topographical map showing the moon's Tycho crater

the fastest and safest way to get there. Without moon maps drawn by men still on earth, there could be no moon exploration.

Mapping the Sky: Weather
9

Perhaps one reason everyone talks about the weather is because it is ever changing. Within a 24-hour period it is not uncommon to have rain, sun, wind, calm, and a 20-degree change of temperature. For this reason, weather maps are printed daily in many newspapers. Land maps often can be used for many years, pilot charts are issued once a month, but new weather maps are made several times a day.

Weather seemed very mysterious to primitive people. Even during the Middle Ages, great storms were looked upon as punishment for evil deeds, or perhaps the work of a witch. The first attempt at scientific prediction of the weather came as a result of the invention of the *barometer* in 1643 by Evangelista Torricelli, an Italian scientist.

The barometer is a device which measures the pressure of the air. A Torricelli barometer can be made by filling a narrow glass tube about 34 inches long with mercury. The tube is turned upside down and placed in a bowl full of mercury. To be sure that no air can bubble up through the mercury, the mouth of the tube is covered while it is being turned. Under normal weather conditions, the liquid metal in the tube will drop until it stands at about 30 inches above the surface of the mercury that is in the bowl.

When the air is heavier, its pressure on the mercury in the bowl is greater and the mercury is forced higher into the tube. When this happens, we say, "The barometer is rising." When the air is lighter, there is less pressure on the mercury in the bowl and the mercury in the tube drops lower. Then we say, "The barometer is falling." It did not take long to discover that bad weather often accompanies a falling barometer.

Weather is an international problem and men from many countries have added to our knowledge. Professor Brande of Breslau University in Poland made one of the greatest contributions. He decided to make charts showing weather changes over a large part of Europe during the year 1783. Generally, a map-maker draws a map so that he can share his knowledge with others quickly and easily. In this case, the map-maker himself gained a great deal of knowledge by drawing his map.

There was no radio, telephone, or telegraph to help him, so Brande slowly collected his information by mail. As he drew the charts, he noticed that local storms were often part of larger storm systems. Many of these storm systems followed the same general path as they moved across Europe.

The knowledge gained from these early charts was more than just an interesting bit of information. It showed that if the path of storm systems is known, weather can be forecast. But weather predictions had to wait until a quicker way to send news was invented. After all, by the time stage coach mail brought the observations, the storm had already come and gone.

MAPPING THE SKY: WEATHER

The invention of the telegraph by Samuel Morse made possible the rapid collection of weather information. When this information was mapped, the course of severe storms could be predicted early enough to send out storm warnings. This service was so valuable that a Federal Weather Bureau was started in 1870.

At first, information gathered by the Weather Bureau was of limited use to ships at sea. Display stations were set up along the coast lines to warn passing ships with a system of flags and lights. This method is still used today, but it is of value mainly to small ships that do not have radios. It is not unusual for a radio weather report to include such information as, "Small craft warnings are being displayed from Cape Hatteras to Block Island."

The invention of the radio made it possible for ships far at sea to receive storm warnings. Also, they could send out information about weather conditions at their position on the ocean. The first weather observation from a ship at sea was sent by the *S.S. New York* in 1905. Its location at that time was 40° N. Lat. and 60° W. Long.

Today a vast network of weather-observation stations circles the earth. From Station Alert at Dumb Bell Bay on Ellesmere Island, which is only 518 miles from the North Pole, from McMurdo Station in Antarctica, from U.S. Coast Guard-maintained ship stations at sea, from weather satellites orbiting the earth, from amateur observers in small towns, the weather news comes flooding in. After this information is mapped, the weather man is ready to make his predictions.

These predictions are sometimes of great help to you in planning a trip or a picnic, but you may be sure that the

UNDERSTANDING MAPS

U.S. Congress does not vote to spend millions of dollars just to make your picnic a success. Weather charts are among the most important maps of the world. The safety of air travel depends on these maps. For example, weather charts were among the maps Lindbergh used in planning his flight across the Atlantic. These showed the type of weather he could expect while crossing. They charted the prevailing winds and thus helped him decide how much gasoline he would need.

Even with today's modern jets, weather maps are still used for the same purposes. Weather maps are used to prepare flight plans that avoid local disturbances such as thunderstorms. Jets flying at high altitudes use a *jet stream* the way ships sailing in the Atlantic Ocean use the Gulf Stream. A jet steam is a strong current of air at about 30,000 to 40,000 feet above the surface of the earth, and it sometimes moves faster than 100 miles an hour. Jet streams move from west to east in the Northern Hemisphere so planes traveling in that direction try to ride a jet stream, while those headed in the opposite direction avoid it. A plane riding a jet stream from the Pacific coast to the Atlantic coast may save as much as one hour in time and from three to six tons of fuel.

The positions of jet streams are not stationary. Jet streams are pushed around by the front edges of cold masses of air flowing down from the polar regions. The advancing edge of a polar air mass is marked on the weather map as a *cold front*. In fact, very often the position of the cold front and the position of the jet stream are similar. Thus, maps showing the positions of the cold fronts are essential to planning long-distance flights, and these same maps are also essential to forecasting the weather.

Mapping and weather forecasting are so closely related that, in 1965, the United States Department of Commerce combined the Coast and Geodetic Survey and the Weather Bureau into the *Environmental Science Services Administration*. Many reports of new discoveries and improvements in mapping, navigation, and weather forecasting are the work of this combined group, which the newspapers refer to as *E.S.S.A.* One of the weather satellites launched in 1966 was called *Essa I.*

The combination of weather satellites and computers has resulted in one of the most exciting break-throughs in weather forecasting since man discovered that weather patterns generally move from west to east in the northern hemisphere. From data relayed by weather satellites to the earth, the weather forecaster can get photographs far enough above the earth to see the cloud formations over tremendous areas. He can watch what happens to the clouds as they move across the surface of the earth. By using computers, he can study the information from the satellites and combine it with information from ground stations. The computers can quickly pick out important trends and predict weather patterns and changes. The computers print out their findings in the form of tables and maps. And slowly, as the information builds up, man is really beginning to understand the operation of the tremendous and mysterious forces which control the weather.

When one of the early satellites was sent up, a scientist complained humorously that it was performing too well. He was being drowned in information. This is rather like complaining that one has received too many birthday presents and does not know which to open first. In time, the computers will fit the bits of information into place, and, while man may never be able to control the weather completely, he will be better able to adapt to it because he will understand what is causing it and so be able to predict it more accurately.

Even though such violent storms as hurricanes cannot yet be controlled, weather maps have helped to save thousands of lives. By plotting the probable course of a hurricane on a map, warnings can be given. These warnings allow time to remove people, animals, and valuables from dangerous places. The story of three hurricanes shows the importance of storm warnings. In October, 1893, an unannounced hurricane swept into New Orleans, bringing with it a great tidal wave and killing about 2,000 people. In September, 1947, the eye of a hurricane passed over New Orleans again. Although there was millions of dollars in property damage, this time, because of early warning, only 12 lives were lost. But hurricanes are still dangerous, and storm warnings alone cannot save lives. People must save themselves by obeying the instructions that are broadcast with these warnings. When hurricane **AUDREY** passed over the bayou section of Louisiana and parts of Texas in June, 1957, 430 people were killed. Many of the deaths occurred because the people did not take the warnings seriously and stayed in their homes until it was too late.

Not all weather-warning services are as dramatic as the hurricane service. Fruit farmers, for example, are also dependent on forecasts. Fruit may freeze if the tempera-

MAPPING THE SKY: WEATHER

ture drops below 32°F., and so the fruit growers watch the weather maps carefully. Many have air-heating devices ready in the orchards to protect the fruit if the temperature goes down to the danger point. Summer hailstorms can also destroy whole crops. In some parts of the country, farmers have banded together to set up cloud-seeding services. When a hailstorm warning is received, a plane flies through the clouds and seeds them with chemicals in the hope that they will drop their moisture as rain rather than as hail. Occasionally the operation succeeds.

Weather maps may even determine whether or not you will have a day off from school. Principals and superintendents depend on weather forecasts to help them decide if the snowfall or sleet will be bad enough to close school.

Weather forecasters use many kinds of maps. Separate maps are made to show the differences in air pressure at various altitudes. Those for the surface level and the 10,000-foot level are most common. Drawn on these maps are *isobars* (eye′so·bars) which are lines showing where the air pressure is the same.

Charts showing temperatures are also made. *Isotherms* (eye′so·therms) are drawn on these maps. An isotherm is a line running through places which have the same temperature. In studying science, you will often find words beginning with "iso-." This comes from a Greek word meaning equal. Whenever you see "iso-" as part of a word, you will know that it means "the same."

Winds Aloft Charts, with arrows to show the direction of the winds, are much used by both weather forecasters and airplane navigators. These are made for such levels as 4,000 feet, 6,000 feet, 10,000 feet, 14,000 feet, etc., above sea level. The weather observer sends up small balloons. From their drift, he can tell the wind direction at the higher levels. He also uses *radiosondes* (ray′dee·o·sonds) to find out more about the upper levels of the air. A radiosonde is a box of weather instruments attached to a balloon. In the box there are a thermometer to measure temperature, a barometer to measure air pressure, a hygrometer to measure moisture, and a small radio sending set. This uses a code of dots and dashes to send the information collected by the instruments back to the weather observer on the ground.

Climate maps are used to show average weather conditions. *Mean* is another word for average, and you can find charts showing monthly and yearly mean rainfall and mean temperatures. One of the first sheets of the new *National Atlas* was the *Mean Monthly Sunshine* for the United States published in 1967 by the U.S. Geological Survey. Climate maps are helpful not only to weather forecasters, but also to people who are planning to start a new industry or to plant a new crop. In an area with a very high average rainfall, no one would want to start a factory where much material had to be stored out-of-doors. By comparing climate charts for many years, it also is possible to see if the climate of a place is changing.

The weather charts you will see most often are the surface weather maps. These are the ones you can find in the newspapers. Surface weather maps show the position of air masses. If you were to visit the North Pole, you would expect to find very cold air. Along the equator you would expect warm air. Just as the water in the ocean is constantly moving, the air too is always in motion. Great masses of cold air from the polar regions of Canada sweep down across the North American continent. Warm, moist air flows northward from the Gulf of Mexico. A *warm front* is the advancing side of a warm air mass. The weather changes when a new front moves in.

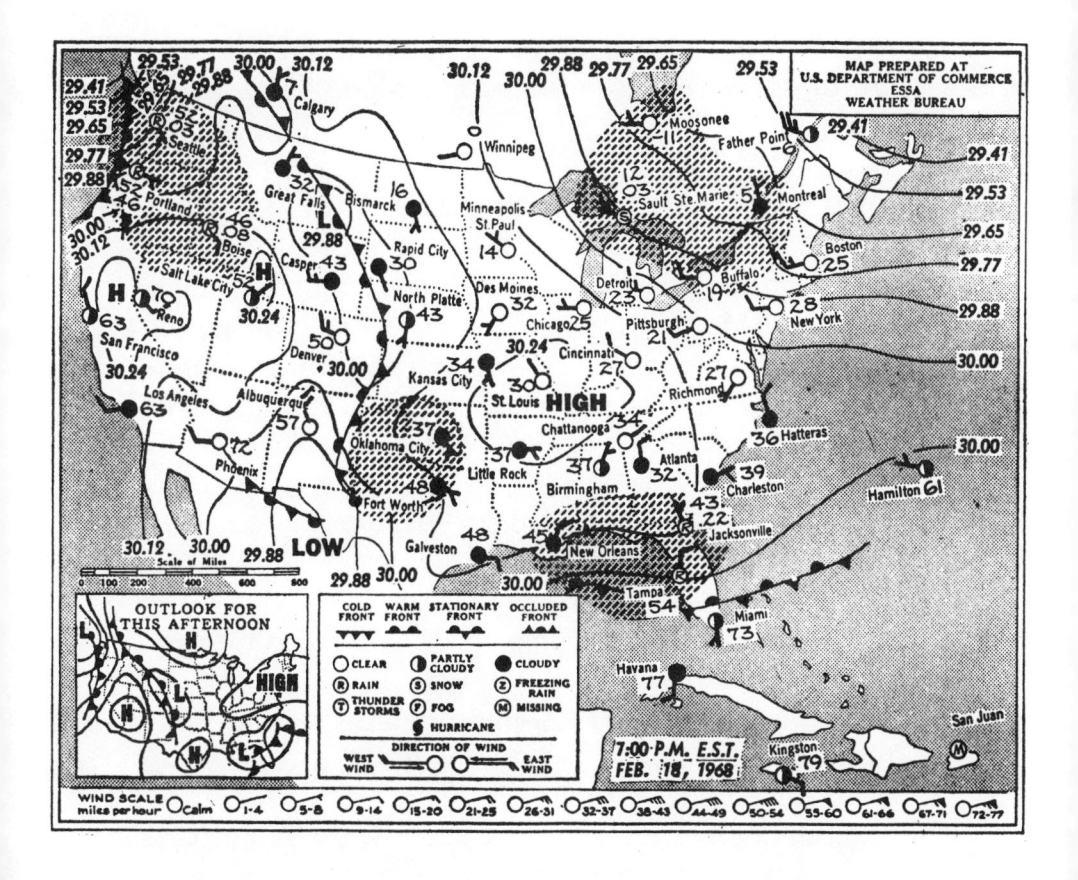

Surface weather maps are shaded to show rain and snow. There are special symbols to mark thunderstorms and hurricanes. Sometimes the temperatures at important weather stations are printed on the map. Some surface

131

UNDERSTANDING MAPS

weather maps show areas of high and low pressure. Isobars are drawn on such maps to show where the pressure is the same.

NOW TRY THIS

Try making your own predictions by using daily weather charts taken from a newspaper. You will need to collect maps for several days so that you will have some idea how fast the air masses are moving. In making predictions you will need to remember the following weather facts:

1. Weather in the Northern Hemisphere generally moves from west to east. Normally it moves about 600 miles in 24 hours.

2. Cold fronts and high-pressure areas tend to move southeast.

3. Warm fronts and low-pressure areas tend to move northeast.

4. Areas of low pressure usually include storms.

5. Areas of high pressure usually have fair weather.

6. Weather changes when a new front moves in.

It is fun to see if your predictions come true. Probably you will be right only part of the time since even the professional forecasters are sometimes wrong. But you can be sure that the most learned scientist 100 years ago could not predict the weather as accurately as you can.

Mapping the Round Earth
10

Through the ages, scientists and other wise men have learned many facts about the earth. Often they have shared their knowledge by picturing it on maps. Sometimes they have gained knowledge by drawing maps. But one thing they have never learned is how to make a flat map of the earth tell the whole truth. Flat maps can show us many important facts, but no matter how hard the mapmaker tries, a flat map of a round world will always show some things which are not true.

Scientists have learned the secrets of atomic energy. Aviators can fly faster than sound, and space ships can be sent beyond the earth's atmosphere. You can send your voice winging over the ocean by radio telephone, but no one can flatten out a globe without tearing, stretching, or crumpling it.

Four things would be true about a perfect flat map of the world: the shapes of the land masses and the oceans would be correct; their sizes would be accurate; it would be possible to find the exact distance between any two places on the map by using the scale; and each place on the map would be found in the right direction from every other place. Every flat map of the earth you have ever seen has been distorted in one way or another.

UNDERSTANDING MAPS

NOW TRY THIS

You can use a baseball to understand some of the mapmakers' problems. Cut a circle out of tissue paper. If you are using a hard ball, the circle should be 9 inches in *diameter*. (The distance across the center of a circle or through the center of a sphere is its diameter.) Cover the baseball as carefully as you can with the paper. Find the sewing lines on the ball and trace them with a pencil. Unwrap and smooth out the paper. You will find that the

FLAT MAP OF A BASEBALL

lines are not connected. Use a colored pencil to join the broken lines. This new line is longer and its shape is different from the original line on your baseball. This is what would happen if a map-maker tried to make a flat picture of a globe in this way.

If globes are more truthful, why bother with flat maps? Imagine the chart room of a yacht about to sail from Boston for Cocos Island, off the west coast of Panama. It would need a globe to show prevailing winds, a globe to show ocean depths and currents, a globe to be used with radio bearings, celestial globes to show the heavens in both the Northern and Southern Hemispheres, and finally, a globe big enough to show the coast line of Cocos Island.

Since the whole island is only 15 miles long, a globe the size of the famous Langlois Globe of Paris, the scale of which is about 5 miles to 1 inch, would be needed. The distance around the equator on the Langlois Globe is 410 feet. The diameter is 128 feet. A globe this size would cover the whole deck of a ship like the *Mayflower,* and hang over 13 feet at both the prow and the stern. Of course, the overhang on each side would be much greater.

Perhaps you would rather imagine your family car ready for a trip across the state. Fastened to the roof would be a global road map about the size of the two-ton aluminum globe in the Daily News Building in New York City. This one has a scale of 55 miles to 1 inch, and the distance around its equator, representing almost 25,000 miles, is only about 38 feet! The flat road maps which you can fold up and keep in the glove compartment of your car usually have a scale of from 10 to no more than 20 miles to 1 inch for a single state.

Obviously, flat maps are needed not only for convenience, but also to show the details of small areas. Flat maps of small areas can be made accurately. If you want to trace on tissue paper an inch or two of the sewing line on your baseball, you can do it easily and almost exactly. The earth is so tremendous that the curve at any one place is very slight. For this reason, flat maps which show only a city, a state, or a small country are accurate enough for general purposes. The real trouble arises when the map-maker tries to show a continent, a hemisphere, or the entire earth on a flat map.

Of course, no map is really made by wrapping a piece of tissue around a globe as you did with the baseball. A map-maker uses geometry to help him make flat maps of the round world. When you show slides or movie films on a screen, you are projecting them. When a map-maker shows a round globe on a flat surface, he makes a *projection* (pro·jec'shun) of the globe. Different kinds of projections are made for different purposes.

Most maps of large areas of the earth are projected in one of three ways: onto a cylinder; onto a cone; or onto a flat or plane surface. A flat map is made from a cylinder or cone by splitting it along a line and opening it out. A map made from one of these projections will be accurate only where the projection touches the globe, but the areas close to this point of contact also will be reasonably accurate.

The cylinders used touch the globe along the equator, so cylindrical projections are accurate for the equatorial regions. A cone generally touches the globe at the 40th

PROJECTIONS

CYLINDRICAL   CONIC   PLANE

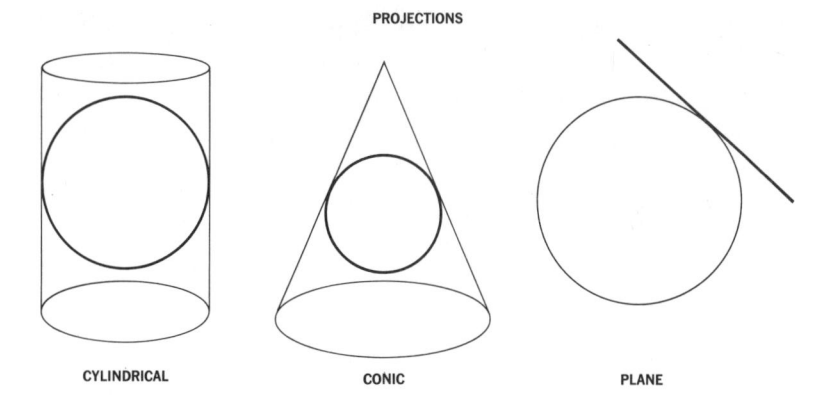

parallel. Since this parallel passes through the United States, a conic projection of the Northern Hemisphere can be used to map the United States.

A plane projection is sometimes called a *polar projection*, but the touching point does not need to be one of the poles. *Polar cases,* which are plane projections touching one of the poles, are very common. But a map can be made which touches the globe at any place: for example, your home town. Again, the map is accurate for the touching point and reasonably accurate for a circle around it.

In the following diagrams a picture of a dog drawn on a globe has been projected by three different methods. As you can see, each of these projections has its faults.

Map-makers have tried to correct these distortions by changing the projections to suit their needs. Gerardus Mercator, a Dutch map-maker who lived in the 1500s, made the best-known variation of the cylindrical projection. The Mercator projection is most important to navigators.

UNDERSTANDING MAPS

As long as ships stayed close to the shores of Europe, the curve of the earth did not matter much to the chartmakers. *Portolan charts* which were made and used by the sailors of the Middle Ages are an example. These maps are covered with straight lines which follow the directions marked on the compass rose. A line which follows one direction all the way is called a *rhumb* (rum) *line*. On a map of a small area, a rhumb line appears to be straight. It is always straight on a Mercator projection, but on all other projections of the globe, a long rhumb line is a curve.

The rhumb-line course of a ship sailing North East by

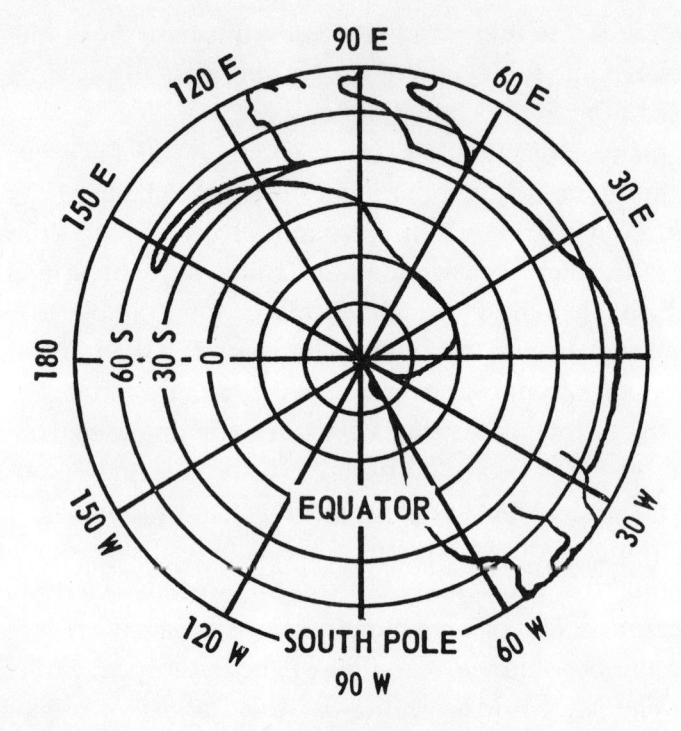

Polar Case Projection

CYLINDRICAL PROJECTION

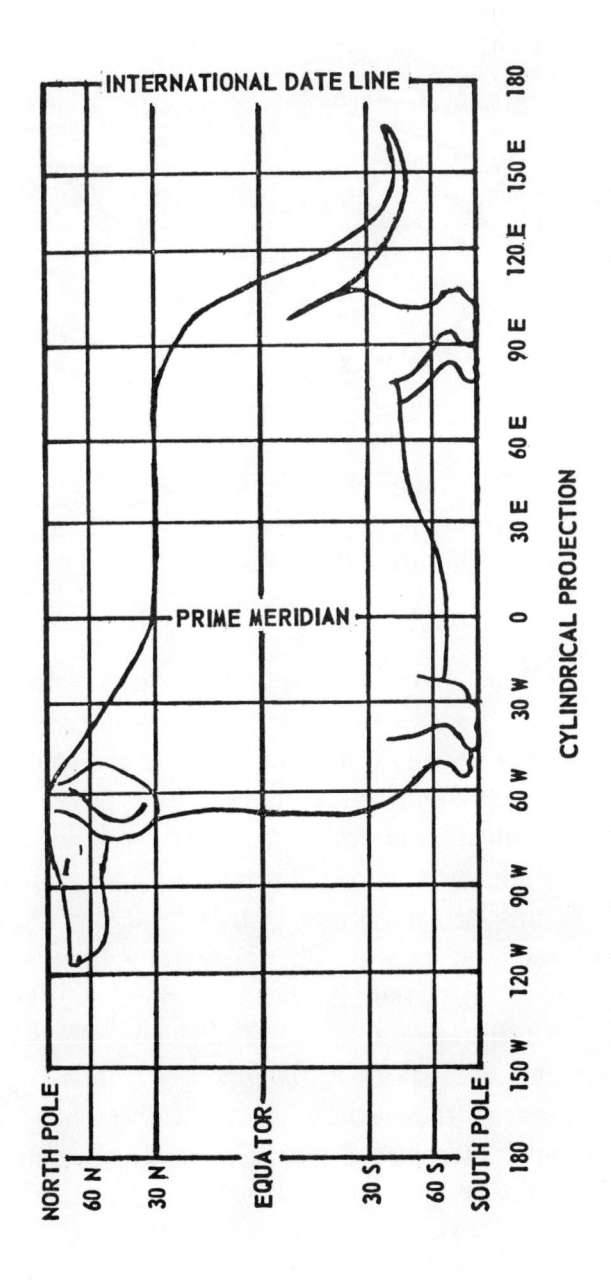

CONIC PROJECTION

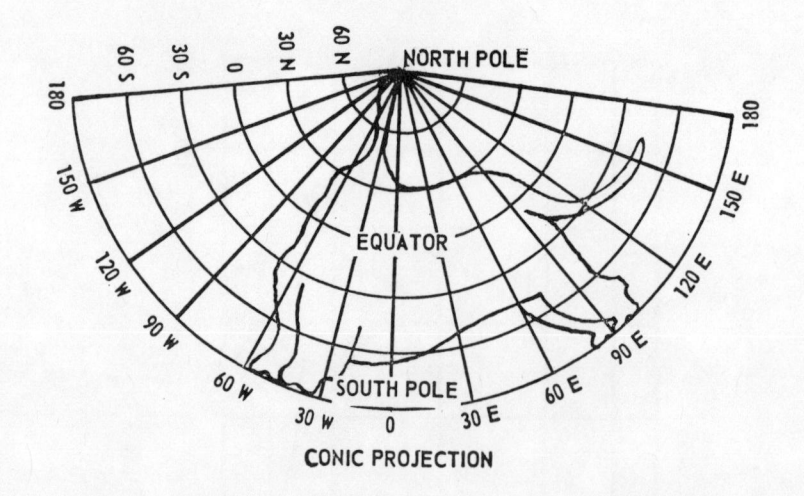

North from the mouth of the Amazon River on the equator (0° Lat. and 50° W. Long.) to Land's End, England (49° 50′ N. Lat. and 6°27′ W. Long.) is plotted on each of the four projections in the following illustrations. You can see that it is a straight line only on the Mercator projection.

The easiest course for a sailor to follow is a rhumb line. His compass direction will be the same for the entire voyage, but he must allow for drift caused by winds and sea currents. Land's End is due North East by North from the mouth of the Amazon River and there are no islands or other obstacles in the way. Therefore, it might seem that a rhumb-line course would be the shortest distance between the two, but this is not true. If the earth were flat, a rhumb line would be the shortest distance between two points. But the curve of the earth, which makes so much trouble for the map-maker, also plays tricks on the navigator. It is about 275 miles farther from the Amazon to Land's End by rhumb line than by a *great circle* route.

140

LANDS END 40 50′ N LAT.
6 27′ W LONG

46 30′ N
37 N
25 30′ N
13 N

50 N, 40 N, 30 N, 20 N, 10 N

PRIME MERIDIAN

EQUATOR

50 W 40 W 30 W 20 W 10 W

AMAZON RIVER

NE BY N RHUMB LINE ON MERCATOR PROJECTION

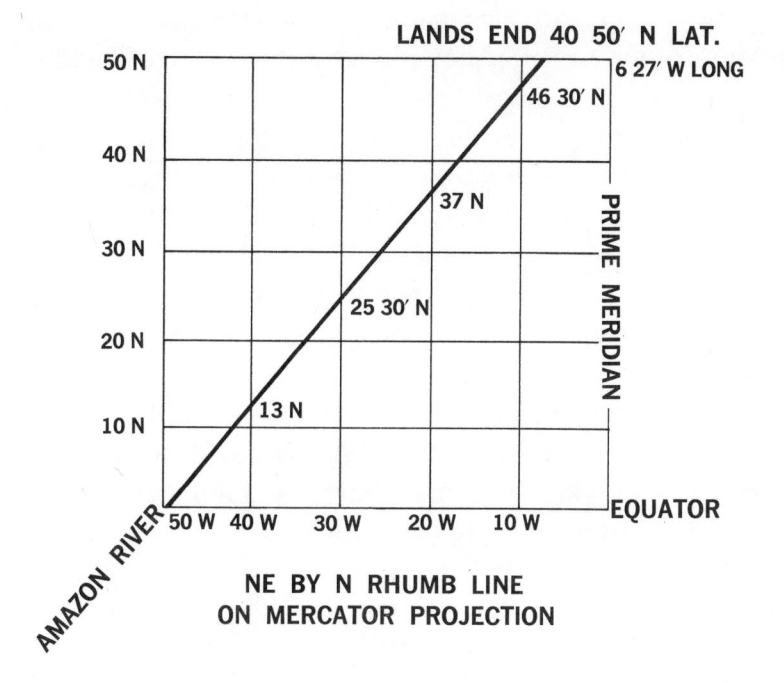

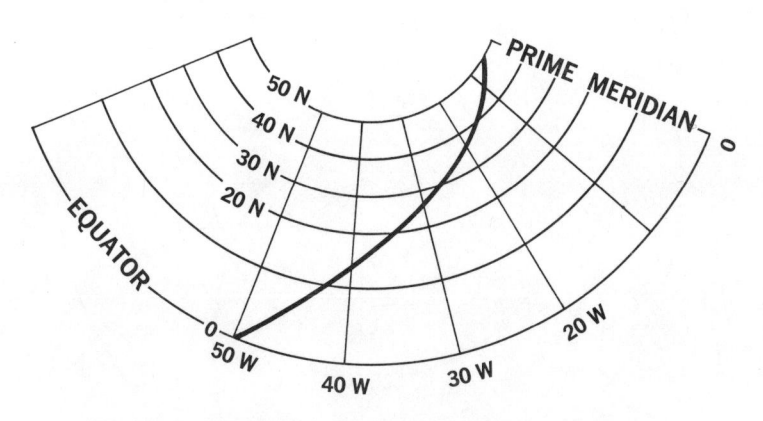

NE BY N RHUMB LINE ON CONIC PROJECTION

141

It is possible to draw many circles on the face of a globe. Some of the circles go completely around the earth and divide it into two equal parts. These are called great circles. There are also *lesser circles,* such as the parallels of latitude (except the equator) which do not divide the globe into equal parts. (See page 68.) The equator and the meridian circles are good examples of great circles, but a great circle can be drawn so that it passes through any two places on the globe. The shortest route between these two places will follow this great circle.

Making the map in sections is still another way of projecting the round earth onto a flat surface. Most globes are made of sections which are printed on flat pieces of paper and then stretched and glued to a sphere.

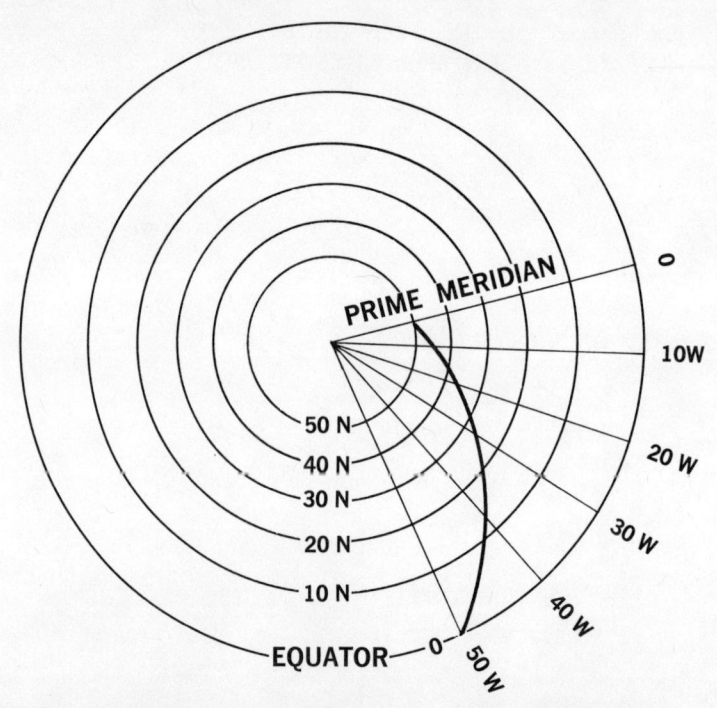

NE BY N RHUMB LINE ON POLAR CASE PROJECTION

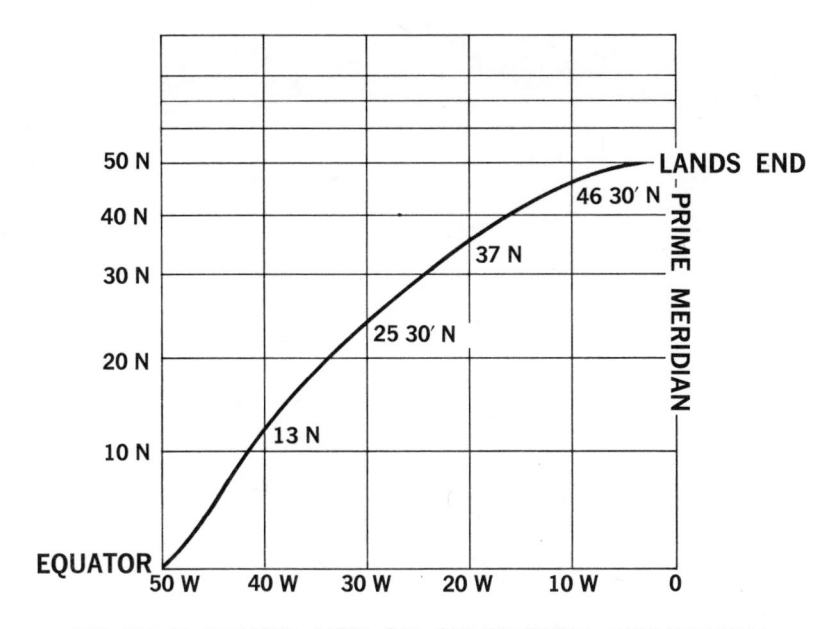

NE BY N RHUMB LINE ON CYLINDRICAL PROJECTION

Sometimes map-makers "peel" the globe and make maps which have curious shapes like the one in the next illustration. These are called *interrupted maps* because they show blank spaces between the parts or *gores* of the maps.

Because maps of small areas are less distorted and show more details than maps of large areas, many map-makers prefer to chart the earth in small sections and then group them in an *atlas*. In the legends of the ancient Greeks, Atlas is the giant who carries the entire world on his shoulders. When Mercator published an atlas of maps bound in one book, he included a family tree showing the ancestors of Atlas. His "atlas" was so popular that soon any collection of maps bound in one book came to be called an atlas.

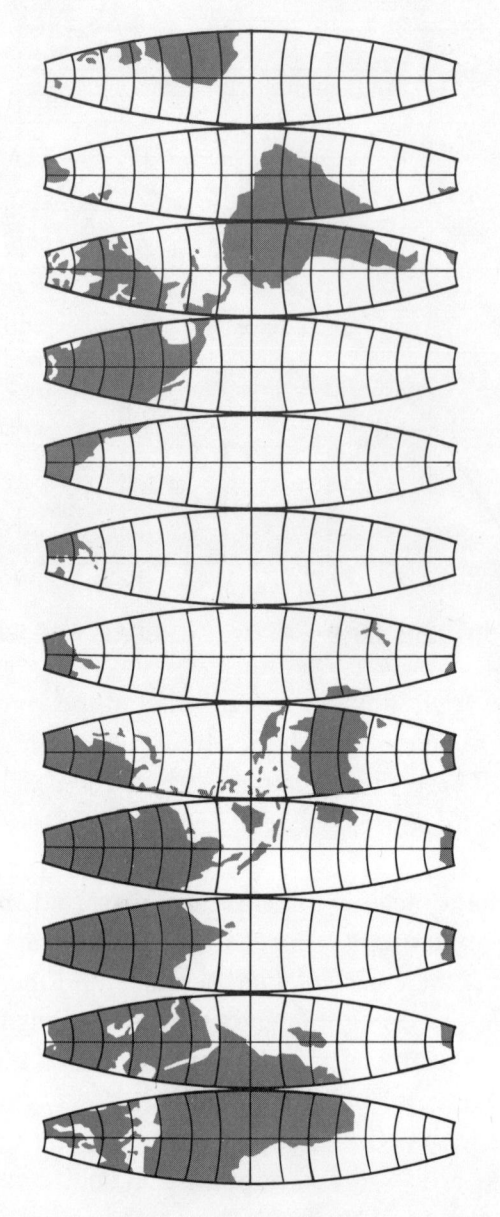

12 GORE MAP

But Mercator was not the first map-maker to make such a collection.

The first known atlas was prepared by Claudius Ptolemy of Alexandria, who lived from 90 A.D. to 168 A.D. It was part of a set of books called *Geographia.* The finest medieval atlas is the *Catalan Atlas,* which can be seen in the Bibliothèque Nationale (National Library) in Paris. It contains Portolan charts of Europe and maps of Asia made from Marco Polo's descriptions of his travels.

The many modern atlases which are published today are the dictionaries and encyclopedias of the map-making world. The U.S. Geological Survey is engaged in preparing a 425-sheet *National Atlas* of which the *Mean Monthly Sunshine Map* published in 1967 is the first sheet. (See page 130.) And the 2,000-sheet atlas of the entire world (see page 35), which was begun more than 50 years ago, is still being compiled.

Mapping the Make-believe, the Almost Real, and the Real
11

Some maps belong to the just-for-fun group. Run your finger down the map of the coast line of North America and then trace the shapes of the islands of the Caribbean. You have just touched the hiding places of billions of dollars of treasure. Somewhere along these coasts lie the loot of pirates and the treasures of water-logged wrecked ships. Key West, Catalina Island, Gardiners Island, Plum Point, and Oak Island—all these are magic names to the seekers of lost treasure.

The search for hidden treasure has attracted men from earliest times. Rich man, poor man, beggar man, thief, everyone loves to dream of finding buried treasure. In July, 1934, the President of the United States, traveling from Puerto Rico to Hawaii, went ashore on a tiny island off the west coast of Panama. Why should a man as busy as President Franklin Roosevelt take time to visit an island inhabited only by goats? He did it because it is a very special island, Cocos, the burial place of three great treasures.

Legend tells us that the first treasure was left there by the English pirate, Edgar Davis, who "banked" his loot on

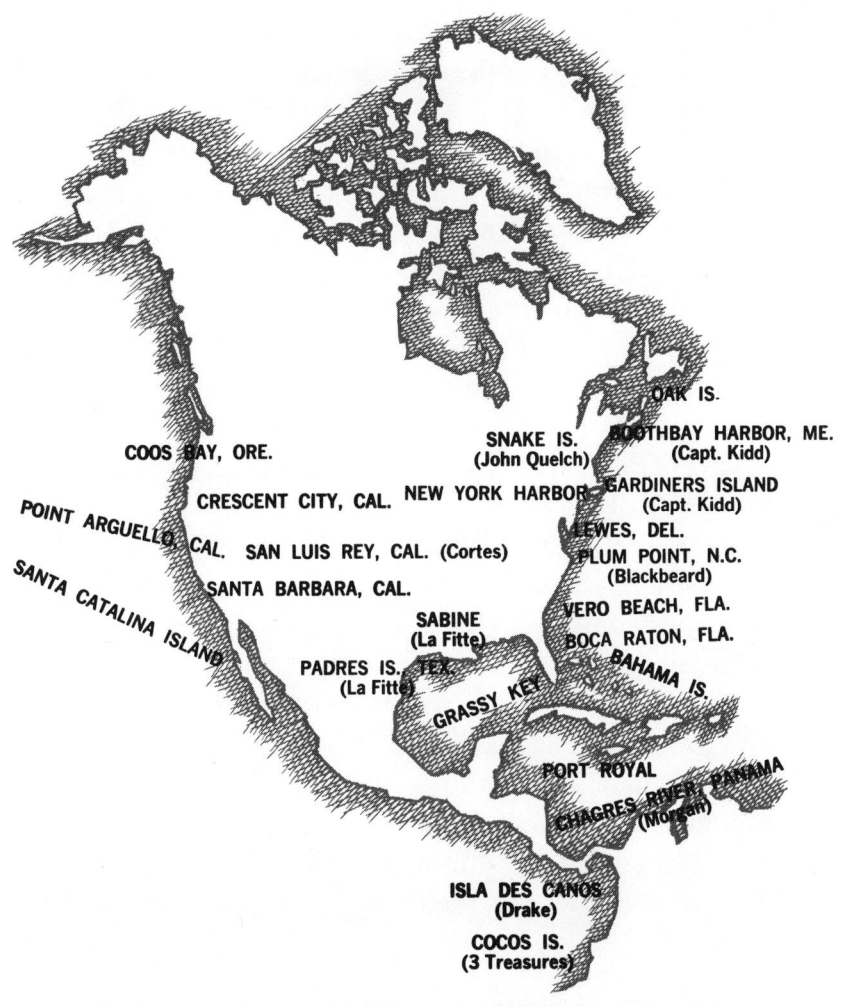

MAP OF A FEW OF THE MANY LEGENDARY TREASURES

those lonely shores. Then, during a revolution in Peru in 1812, the treasures of the cathedral of Lima were said to have been given into the care of Captain Thompson of the ship *Mary Dyer*. But things were so bad in Peru that he decided to bury them on the north end of Cocos Island until he could return them to the rightful owners. Seven years later, the Mexican bandit, Benito Bonito, chose the same island to hide millions of dollars of silver which he had stolen from the Mexican treasure trains. These three stories have attracted many parties of treasure hunters, including the President of the United States. But so far, only one silver crucifix has been found, and the goats of Cocos Island are still the richest goats in the world.

Since Cocos Island is so very small, less than 15 square miles, why has no one discovered any treasure? If you were to visit this rocky island with its waterfalls, caves, and sandy beaches, you would see that it is impossible to dig up the entire place. To find treasure, you would need a map showing where it is located. No real treasure maps of Cocos Island are known to exist, but perhaps somewhere there may be a well-worn chart waiting to be found. Certainly, many false treasure maps of Cocos have come to light. After the silver crucifix was discovered in 1903, false maps were made and sold in England. These mapmakers had found a way to collect treasure without ever setting foot on Cocos Island.

Cocos Island is not the only place that has attracted false map-makers. Maps claiming to show the hiding places of the treasures of Captain Kidd, Blackbeard, Henry Morgan, and others appear all the time. Perhaps no real pirate maps do exist, but we do know that the pirates who sailed the seas so many years ago were unable to keep their riches in banks. So some loot was buried on uninhabited islands. It is easy to see how the legends of pirate maps grew. Certainly, if such maps ever existed, they must have been very poor. With only a compass to help him, the pirate would have had to depend mainly on landmarks, and most landmarks change with the passage of time.

The Library of Congress in Washington has many maps and books which describe locations where lost, buried, or sunken treasures are thought to exist. If you are interested, you can write to the Superintendent of Documents, Government Printing Office, Washington, D.C. 20402 for the *Descriptive List of Treasure Maps and Charts in the Library of Congress.* The cost is 30 cents. Neither the Library of Congress nor the Superintendent of Documents sells these maps or charts, but the list does include the names and addresses of the publishers of some of the maps.

Even if you never see a real treasure map, there are plenty of imaginary ones waiting for you between the covers of books. A good example of such a map can be found in Robert Louis Stevenson's book *Treasure Island.* It is believed that Stevenson based his map on a real map of Cocos Island. Look at the next two maps. You can see that their outlines are similar.

Everyone knows that such maps as the one in *Treasure Island* are make-believe. But there are real maps of imaginary places. The makers of these maps honestly believed that the lands which they drew existed. During the fifteenth and sixteenth centuries, when the countries of Europe first started to explore the rest of the world, maps

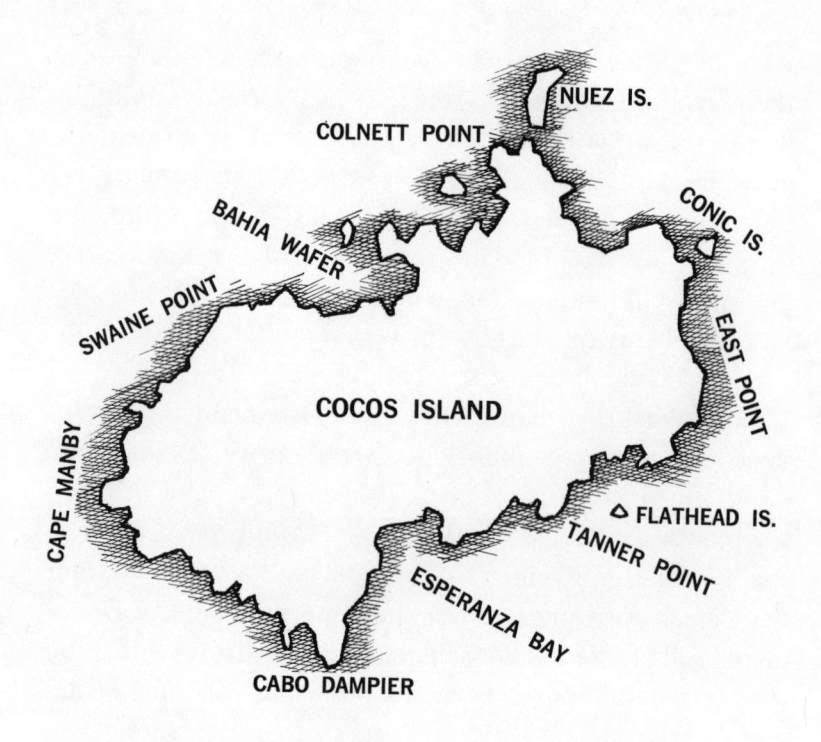

[Map of Cocos Island with labels: NUEZ IS., COLNETT POINT, CONIC IS., BAHIA WAFER, EAST POINT, SWAINE POINT, COCOS ISLAND, CAPE MANBY, FLATHEAD IS., TANNER POINT, ESPERANZA BAY, CABO DAMPIER]

were state secrets, guarded as carefully as today's nuclear bomb secrets. In most countries, it was an act of treason, punishable by death, to sell or give a navigation chart to a foreign nation.

There were two reasons for this. First, since each European country wanted to claim as much new land as possible, you can be sure no country wanted to help its neighbor by sharing its maps. Second, maps were difficult to make, for each one had to be drawn carefully by hand. A map-maker would work a long, long time to prepare charts from the information brought back from one expedition. Some of the travelers who brought back information for the map-makers had very good imaginations and so you

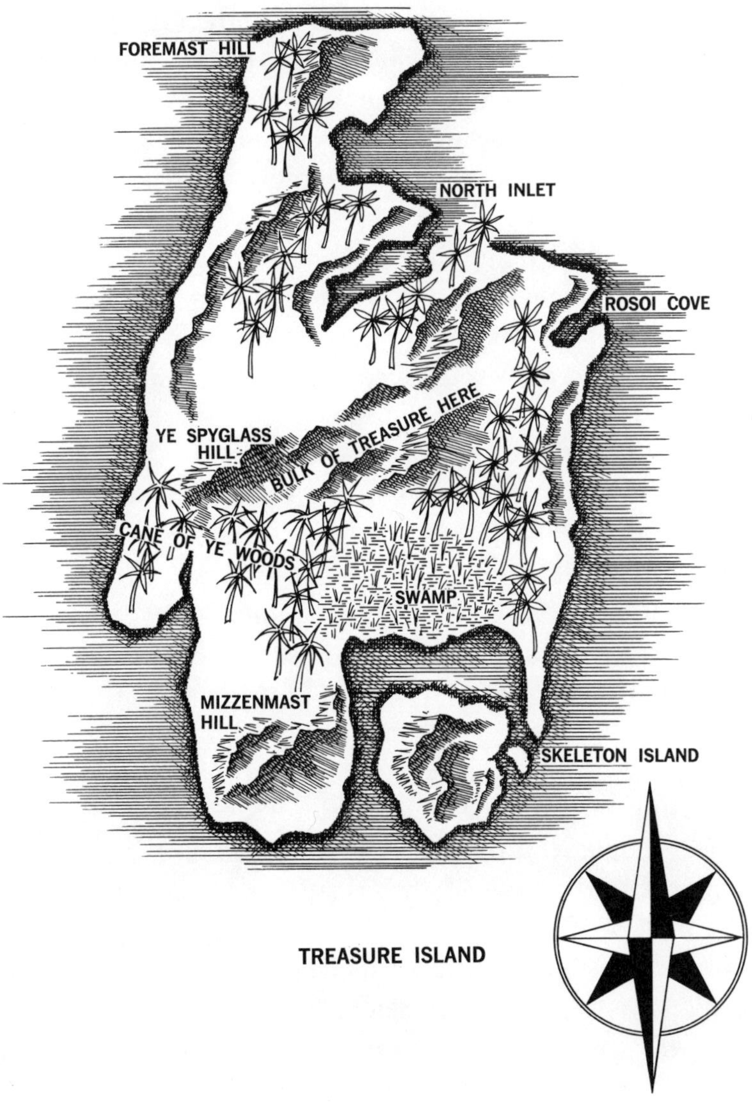
FOREMAST HILL
NORTH INLET
ROSOI COVE
YE SPYGLASS HILL
BULK OF TREASURE HERE
CANE OF YE WOODS
SWAMP
MIZZENMAST HILL
SKELETON ISLAND
TREASURE ISLAND

will find "Isles of Satan," "Isles of Goats," and many other strange islands dotting the North Atlantic Ocean.

One such legendary island was named Buss. It was "sighted" first about 1578, somewhat south of Greenland. It was reported to be one of the largest islands on earth and many expeditions set out to explore it. Although they searched carefully, the nearest thing to an island that ever appeared was an upturned boat bottom covered with moss. But the map-makers hated to give up the island and so Buss appears on maps made as late as 1745. By that time, however, it was reduced to a "surf" ¼ mile long surrounded by a rough sea.

The most famous lost island is Atlantis. The story of this tremendous continent in the middle of the Atlantic originated about 2,300 years ago. It is described as being unbelievably wealthy, with cities and streets of gold. Plato, the Greek philosopher, claimed that "There occurred violent earthquakes and floods, and in a single day and night . . . the island of Atlantis disappeared in the depths of the sea. . . ." Plato located Atlantis beyond the Pillars of Hercules (the Strait of Gibraltar). Old maps sometimes show Atlantis located in the Sargasso Sea.

Although it supposedly disappeared into the ocean many centuries ago, Atlantis is still a favorite location for the writers of imaginative books and movies, and for scientific investigation. In 1967, geologists found volcanic ash in the Mediterranean Sea off the coast of Greece. The ash was from a volcano which erupted about 1400 B.C. Excavations of the ocean floor in that area have uncovered a large town that was destroyed by the volcano. Whether or not this town was located on Atlantis will undoubtedly be the subject of long arguments for many years to come.

Old maps, themselves, may be treasures. Many people collect old maps and are willing to pay large sums for unusual maps. The interrupted map on page 144 may look quite modern to some people. However, an interrupted map of the world, drawn in 1507, was sold in 1960 for $35,000. It is considered especially valuable because it was one of the first two maps showing the name *America* on the Western Hemisphere.

Perhaps the most valuable treasure maps are the maps which show where oil, gold, or uranium may be found. If it is impossible to find the treasures of Cocos Island by digging up a mere 15 square miles, imagine how hard it would be to find the treasures of the earth if prospectors had to dig up the entire 9 million square miles of North America. Again, maps are needed to pinpoint the treasure. But the modern maps differ from the older ones. There is no "X" to mark the spot where the treasure is hidden. Instead, these maps show the kinds of rock that probably lie under the surface of the earth. With this information, a geologist can guess where such modern treasures as oil, natural gas, or uranium may be found.

Modern treasure maps may be prepared by private companies. In this case, they are guarded almost as carefully as were the maps of fifteenth-century Europe. Oil maps stolen in Pittsburgh, Pennsylvania, in 1956, were said to be valued at over a million dollars. Fortunately for the oil company, the maps were recovered.

The U.S. Geological Survey prepares similar maps which are available to everyone. A map which shows the rock formation of the land is called a *geologic map*.

UNDERSTANDING MAPS

Geologic maps of many sections of the United States may be bought from the Geological Survey in Washington, D.C. One such geologic map, printed below, shows a rock formation in New York State in which oil might be found.

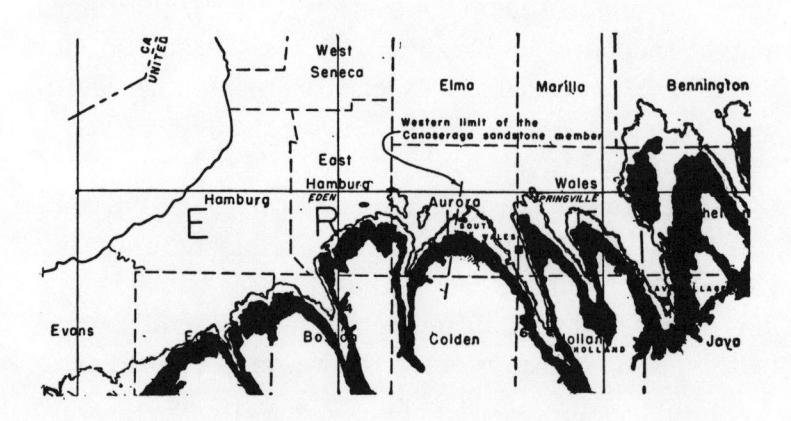

There are other types of maps which are of great importance to you and your family. The oldest and perhaps the most valuable of these are the *land use maps*. These are maps of small areas which show the division, ownership, and use of the land. The oldest existing land use maps were made on clay tablets in Babylon over 4,000 years ago. The newest ones are still on the drawing boards of the map-makers. There is a map on file in your County Courthouse showing the lot on which your house is built.

Another type of land use map has become important recently. For many years, cities just grew. People put up houses and made streets wherever they wanted. One famous exception to this practice is Washington, D.C. It was planned by the French architect, Pierre l'Enfant, who used a system of circles and spokes spreading out from the

Capitol. But beyond the area of his original map, even Washington has grown without a plan. Today we know it is important to make plans of our cities to show where homes can be built, where to locate a shopping center, and where factories can be set up. This is called zoning, and most cities and towns today have *zoning maps*.

Both the state and federal governments issue many different kinds of maps. There are maps which show the voting districts of an area so that each citizen will know where to vote. There are maps of the forest preserves; the Army and the Navy have many, many maps; there are maps showing the density of population across our country. In fact, there is no department in our government which does not use maps.

You may wonder why they make all these special types of maps. They could write out the information instead of plotting it on a map. But imagine, if you can, how many pages of writing it would take to describe the boundaries of every lot in even a very small village! And yet, it can all be shown very clearly on a single map. There is a famous saying, "One picture is worth a thousand words." And a map is a very useful picture.

Index

aerial photography, 46–48, 49, 60, 104
air-mail service, 109
Alpha mountain range, 120
Andrea Doria, 108
angular measurement (*see* measurement, angular)
Army Corps of Engineers, 60
astronomy, 114, 118
Atlantis, island of, 152–153
atlas, 143–145
axis, of earth, 65, 69, 75

Bache, Alexander D., 102
barometer, 123–124
bench marks, 29–30, 53
Beta mountain range, 120
Big Dipper, 23–24, 115
Blackbeard, 148
Board of Geographic Names, 57
Bonito, Benito, 148
Brande, Professor, 124
Buss, legendary island of, 152

Carpathia, 64
cartographer, 49–60
Catalan Atlas, 145
caterpillar map, 51
Central Time Zone, 91
chronometer, 89, 90, 92
circumference, of earth, 91
Civil War, 45
Clemens, Samuel, 96–97
Cocos Island, 66, 135, 146–149, 153

cold front, 126, 132
color separators, 58
Columbus, Christopher, 11, 12, 84, 114
compass, 24–27, 30, 90, 111, 113, 114, 140, 149
compass rose, 16, 26, 111, 138
computers, 60, 75, 92–93, 120, 127–128
constellations, 115, 117
contact flying, 108, 113
continental drift, 75
contour lines, 52

Davis, Edgar, 146
dead reckoning, 84
declination, 117, 118
dipsey lead, 97
Dubhe (*see* Ursa Major)

ecliptic, 117–118
Egyptians (*see* measurements, systems of)
elevation, 42, 45, 52
English (*see* measurements, systems of)
Environmental Science Services Administration (E.S.S.A.), 127
equator, 65–66, 69, 84, 91, 100, 117, 142

fathogram, 97
fathoms, 96–97
fathometer, 97

157

Federal Weather Bureau, 125, 127
Franklin, Benjamin, 102
French (*see* measurements, systems of)

Galilei, Galileo, 120
geodetic surveyors, 32–34, 36–37
geodimeter, 38–39
Geographia, 145
geologist, 12, 153
geometry, 65, 69, 136
glacier, 94–95, 99
great circles, 140–142
Gulf Stream, 102, 126

hachuring, 51
Hamilton Rice Expedition, 46
hand lead, 96
Harrison, John, 90
horizon, 69–70, 81
hurricanes, 128, 131

icebergs, 61–62, 99, 107
Indian Trigonometric Survey, 31
Instrument Approach and Procedure Charts, 111
International Date Line, 87
International Ice Patrol, 62
interrupted maps, 143
invar tape, 36, 38, 39
isobars, 129, 131
isotherms, 129

jet stream, 126

Kidd, Captain, 148

land use maps, 154
lasers, 39
latitude, 66–67, 69–70, 72–74, 76, 77, 81, 94, 114, 116
L'Enfant, Pierre, 154
lesser circles, 140
Lewis and Clark Expedition, 13
Library of Congress, 149

Lindbergh, Charles, 113, 126
Little Dipper, 23–24
lodestone, 25
longitude, 77–83, 84, 90, 94, 114, 116
loran (LOng RAnge Navigation), 104–106
Lowell, Percival, 118
lunar seas, 120

magnet, 25
magnetite (*see* lodestone)
Marshall Islands, 9–10
Mary Dyer, 148
Mauna Kea, 95
measurement, angular, 64, 71
measurement, systems of,
  Egyptian, 35–36
  English, 18
  French, 17–19
Mercator, Gerardus, 137, 143–145
Mercator projection, 138–140
meridian, 77–83, 91, 142
micro chain, 39
Middle Ages, 25, 109, 123, 138
moon, 120–122
Morgan, Henry, 148
Morse Samuel, 125
mosaic map, 53
Mt. Everest, 31, 32, 42, 49, 95

National Atlas, 130, 145
Nautical Almanac, 70, 73, 83
nautical mile, 113
Naval Observatory, 70
negative engraver (*see* scribers)
North Pole, 65, 91, 93, 117
North Star, 23–24, 67–69, 70, 71, 72, 73, 115–116, 117
Nuremberg Eggs, 84

ocean, boundaries of, 103
  currents of, 101–103, 107, 135
  floor, 95–97, 152
orbit, 93

INDEX

parallels of latitude, 67, 72, 142
photogrammetric strips, 46
photogrammetry, 46, 50, 120
pirates, 146–149
planets, 117
Plato, 152
Polaris (see North Star)
Pole Star (see North Star)
Polo, Marco, 145
Portolan charts, 138
prime meridian, 79–81, 83, 91
projection, 136–137
protractor, 71–72
Ptolemy, Claudius, 145

quadrangles, 32–34, 58–60

Radio Facilities Charts, 111
radiosondes, 130
rhumb line, 138–140
right ascension, 117, 118
Roosevelt, Franklin, 146
Ross Ice Shelf, 61

safeway, 108
satellite, 92–93, 106, 114 (see also weather satellite)
scale, 20
scribers, 57–58
sextant, 70, 71, 83, 88, 92
Solar Time, 81–83, 89, 90
South Pole, 65, 73, 91, 93
Sputnik, 92–93
S.S. *New York*, 125
Standard Civil Time, 86
stars, 114–115
stereo pictures, 53–56
stereoscopic plotter, 53–56
Stockholm, 108
storm systems, 124
Strait of Gibraltar, 101, 152

telegraph, 125
theodolite, 39–40, 42, 45, 60

time zones, 85–89, 114
Titanic, 62–64, 79
Torricelli, Evangelista, 123
Treasure Island, 149
triangulation, 37–42, 60, 92
Twain, Mark (see Clemens, Samuel)

UNAMACE (Universal Map Completion Equipment), 60
United States Coast and Geodetic Survey, 46, 50, 75, 97, 99, 102, 109, 111, 127
United States Geological Survey, 32, 130, 145, 153
United States Hydrographic Office, 62, 99
United States National Geodetic Satellite Program, 92
universal map, 35–36
Ursa Major, 115–116

variation-of-latitude observatories, 74–75
Venus, 119
Vinland Map, 11–12

warm front, 131, 132
Washington, George, 36
Weather Bureau (see Federal Weather Bureau)
weather observation stations, 125, 131
weather satellites, 125, 127
Whistler, James McNeill, 51
World War I, 108
World War II, 32, 35, 49, 101, 104

zenith, 81, 83
zoning maps, 155

159